Contents

CW01501218

CONTENTS

CONTENTS

CONTENTS

CONTENTS

CONTENTS

Preface

This is the seventh edition of Spon's Minor Works, Alterations, Repairs and Maintenance Contractors' Handbook. The book is intended to help the contractor who is on the point of becoming self-employed or has already taken the plunge and is now operating a small contracting business.

Well established contractors can also benefit by using this book, and from the helpful and constructive criticisms received, it would appear they are among the most regular users of the wealth of information contained in the following pages.

Advice on the business side of the industry has been condensed into the first three chapters and is set out in an easy-to-read style with many examples. The information given on current tax rates is taken from the November 1993 budget proposals.

It is not intended that these chapters should replace the need for professional advice when the occasion warrants it but they are meant to complement this need and hopefully save money in consultation fees.

Chapter 4 sets out some of the basic principles of estimating but the main thrust of the book lies in Chapter 5 - Rates for Measured Work. The information contained in this chapter is intended as a base on which an estimator can produce his quotations.

It cannot be stressed too strongly that despite claims made on their behalf, price books should not be used as a literal source of information for preparing quotations, but as a base on which a contractor can formulate his own unique pricing data.

The measured items are presented in accordance with the requirements of the 7th Edition of the Standard Method of Measurement of Building Works (SMM7). Although this may be irrelevant to the needs of the smaller contractors it would be required by those larger firms tendering for work on tender documents based upon SMM7.

The indexing and reference systems have been improved so that the self-employed contractor who prices jobs without formal tender documents can also receive the maximum benefit from the book. It should be noted however that not

PREFACE

all of the requirements of SMM 7 have been observed and it is hoped that the compromise will produce a balance which will suit the majority of readers.

There are many items affecting the value of building works. The speed and skill of individual workmen and the price paid for materials are factors which can directly affect the profitability of a job and the rates in Chapter 5 should only be used as a basis for preparing an estimate or quotation.

These comments do not detract from the real value of the information in this chapter. Over 5,000 item descriptions and unit rates are given which are based on 30,000 separate pieces of data providing a wealth of detailed cost information.

The editors have received a great deal of help from many sources in the research and the preparation of this book and would like to acknowledge the assistance given with grateful thanks. Grant Thornton provided the information and advice on business and taxation matters set out in Chapters 1 to 3.

The editors would welcome constructive criticism of the book together with suggestions for improving its scope and contents. Whilst every effort has been made to ensure the accuracy of the information given in this publication, neither the editors nor the publishers in any way accept liability of any kind resulting from the use made by any person of such information.

The male pronoun is used in this book. This is for ease of reading and should be taken to mean both male and female individuals.

Bryan J.D. Spain FInstCes, MACostE
TWEEDS (incorporating SPAIN AND PARTNERS)
Chartered Quantity Surveyors
Cavern Walks
8 Mathew Street
Liverpool L2 6RE

SMM6/SMM7 Table

The following table is intended to assist the reader identify the new work sections in SMM7 by listing them adjacent to the same elements in SMM6.

SMM6	SMM7
A GENERAL RULES	
B PRELIMINARIES	PRELIMINARIES/GENERAL CONDITIONS
Preliminary Particulars	A10 Project particulars
	A11 Drawings
	A12 The site/existing buildings
	A13 Descriptions of the work
Contract	A20 The contract/sub-contract
	A30 Employer's requirements: tendering/sub-letting/supply
	A31 Employer's requirements: provision, content and use of documents
	A32 Employer's requirements: management of the works
	A33 Employer's requirements: quality standards/control
	A34 Employer's requirements: security/safety/protection
	A35 Employer's requirements: specific limitations on method/sequence/timing

SMM6	SMM7

B PRELIMINARIES (cont'd)

	Specific limitations on method/ sequence/timing
	A36 Employer's requirements: Facilities/Temporary works/Services
	A37 Employer's requirements: Operation/Maintenance of the finished building
Works by Nominated Sub-Contractors	A50 Work/Materials by the Employer
Goods and Materials from	A51 Nominated sub-contractors
Nominated Suppliers and	A52 Nominated suppliers
Works by Public Bodies	A53 Work by statutory authorities
General facilities and obligations	A40 Contractor's general cost items: Management and staff
	A41 Contractor's general cost items: Site accommodation
	A42 Contractor's general cost items: Services and facilities
	A43 Contractor's general cost items: Mechanical plant
	A44 Contractor's general cost items: Temporary works
Contingencies	A54 Provisional work
	A55 Dayworks

C DEMOLITION	C DEMOLITION/ALTERATION/ RENOVATION
Generally	C10 Demolishing structures
	C20 Alterations - spot items
	C30 Shoring
	C40 Repairing/Renovating concrete/ brick/block/stone
	C41 Chemical dpcs to existing walls
	C50 Repairing/Renovating metal
	C51 Repairing/Renovating timber
	C52 Fungus/Beetle eradication
Protection	A42 Contractor's services and facilities

SMM6/SMM7 TABLE

SMM6	SMM7
D EXCAVATION AND EARTHWORK	**D GROUNDWORK**
Site preparation	D10 Ground investigation
	D11 Soil stabilization
Excavation	D12 Site dewatering
	D20 Excavation and filling
for services etc.	P30 Trenches/Pipeways/Pits for buried engineering services
Earthwork support	
Disposal of water	
Disposal of excavated material	
Filling	
Surface treatments	
Protection	A42 Contractor's services and facilities
F CONCRETE WORK	J31 Liquid applied waterproof roof coatings
	J32 Sprayed vapour barriers
	J33 In situ glass reinforced plastics
Damp-proof membranes	J30 Liquid applied tanking/damp-proof membranes
	J40 Flexible sheet tanking/damp-proof membranes
Surface sealers	M60 Painting/Clear finishing
Holes and chases for services	P31 Holes/Chases/Covers/Supports for services
Reinforcement	E30 Reinforcement for in situ concrete
	E31 Post tensioned reinforcement for in situ concrete
Formwork	E20 Formwork for in situ concrete
Precast concrete	
small units	F31 Precast concrete sills/lintels/ copings/features

SMM6	SMM7

F CONCRETE WORK (cont'd)

	H14 Concrete rooflights/pavement lights
large units	E50 Precast concrete large units
	H40 Glass reinforced cement cladding/ features
	H50 Precast concrete slab cladding/ features
Composite construction	E60 Precast/Composite concrete decking
Hollow block suspended construction	-
Prestressed concrete work	-
Contractor-designed construction	-
Protection	A42 Contractor's services and facilities

G   BRICKWORK AND BLOCKWORK   **F   MASONRY**

Brickwork	F10 Brick/Block walling
Brick facework	F10 Brick/Block walling
Brickwork in connection with boilers	-
Blockwork	F10 Brick/Block walling
Glass blockwork	F11 Glass block walling
Damp-proof courses	F30 Accessories/Sundry items for brick/block/stone walling
Sundries	
generally	F30 Accessories/Sundry items for brick/block/stone walling
Bedding and pointing frames	L10-12 Windows
	L20-22 Doors etc.
	F30 Accessories/Sundry items for brick/block/stone walling

SMM6	SMM7
Cavity insulation	P11 Foamed/Fibre/Bead cavity wall insulation
Holes	- deemed included - except P31 Holes/Chases/Covers/Supports for services
Centering	- deemed included
Protection	A42 Contractor's services and facilities

H   UNDERPINNING

Work in all trades	D50 Underpinning
Protection	A42 Contractor's services and facilities

J   RUBBLE WALLING

Stone rubble work	F20 Natural stone rubble walling
Sundries	F30 Accessories/Sundry items
Centering	F20 Natural stone rubble walling
Protection	A42 Contractor's services and facilities

K   MASONRY

Natural stonework	F21 Natural stone/ashlar walling/dressing
	H51 Natural stone slab cladding/features
Cast stonework	F22 Cast stone walling/dressings
	H52 Cast stone slab cladding/features
Clayware work Sundries	F30 Accessories/Sundry items
Holes and chases for services	P31 Holes/Chases/Covers/Supports for services

SMM6	SMM7
K MASONRY (cont'd)	
Other holes, cramps etc.	F21 Natural stone/ashlar walling/dressing F22 Cast stone walling/dressings
Centering	F21 Natural stone/ashlar walling/dressing F22 Cast stone walling/dressings
Protection	A42 Contractor's services and facilities
L ASPHALT WORK	**J WATERPROOFING**
Mastic asphalt	J20 Mastic asphalt tanking/damp-proof membranes J21 Mastic asphalt roofing/insulation/ finishes M11 Mastic asphalt flooring
Asphalt tiling	
Protection	A42 Contractor's services and facilities
M ROOFING	**H CLADDING/COVERING**
Slate or tile roofing	H60 Clay/Concrete roof tiling H61 Fibre cement slating H62 Natural slating H63 Reconstructed stone slating/tiling H64 Timber shingling H76 Fibre bitumen thermoplastic sheet coverings/flashings
Corrugated or troughed sheet roofing or cladding	H30 Fibre cement profiled sheet cladding covering/siding H31 Metal profiled/flat sheet cladding/ covering/siding H32 Plastics profiled sheet cladding/ covering/siding

SMM6	SMM7
	H33 Bitumen and fibre profiled sheet cladding/covering
	H41 Glass reinforced plastics cladding/features
	K12 Under purlin/Inside rail panel linings
	G30 Metal profiled sheet decking
	G31 Prefabricated timber unit decking
Roof decking	G32 Edge supported/Reinforced woodwool slab decking
	J22 Proprietary roof decking with asphalt finish
	J43 Proprietary roof decking with felt finish
	K11 Rigid sheetflooring/sheathing/linings casings
Bitumen-felt roofing	J41 Built-up felt roof coverings
	J42 Single layer plastics roof coverings
Sheet metal roofing	H70 Malleable metal sheet prebonded coverings/cladding
Sheet metal flashings and gutters	
	H71 Lead sheet coverings/flashings
	H72 Aluminium sheet coverings/flashings
	H73 Copper sheet coverings/flashings
	H74 Zinc sheet coverings/flashings
	H75 Stainless steel sheet coverings/flashings
Protection	A42 Contractor's services and facilities
N WOODWORK	G STRUCTURAL/CARCASSING METAL/TIMBER
Carcassing	G20 Carpentry/Timber framing/First fixing

SMM6	SMM7
N WOODWORK (cont'd)	
First fixings boardings, grounds, framework etc.	G20 Carpentry/Timber framing/ First fixing
external weatherboarding	H21 Timber weatherboarding
flooring	K20 Timber board flooring/ sheathing/linings/casings K21 Timber narrow strip flooring/ linings
Second fixings skirtings, architraves etc.	P20 Unframed isolated trims/ skirting/sundry items
sheet linings and castings	K11 Rigid sheet flooring/sheathing/ linings/castings K13 Rigid sheet fine linings/panelling
	L42 Infill panels/sheets
Composite items trussed rafters etc.	G20 Carpentry/Timber framing/ First fixing
	L WINDOWS/DOORS/STAIRS
windows, window frames etc.	L10 Timber windows/rooflights/ screens/louvres
doors, door frames etc. staircases	L20 Timber doors/shutters/hatches L30 Timber stairs/walkways/balustrade
	N FURNITURE/EQUIPMENT
fittings	N10 Fixtures/furnishings/equipment N11 Domestic kitchen fittings N20, N21, N22, N23 Special purpose fixtures/furnishings/equipment

SMM6/SMM7 TABLE

SMM6	SMM7
Sundries	**P BUILDING FABRIC SUNDRIES**
plugging	deemed included with fixed items (however rates retained under G20)
holes in timber	deemed included - except P31 Holes/Chases/Covers/Supports for services
insulating materials	P10 Sundry insulation/proofing work/fire stops
metalwork	G20 Carpentry/Timber framing/First fixing
Ironmongery	N15 Signs/Notices
	P21 Ironmongery
Protection	A42 Contractor's services and facilities
P STRUCTURAL STEELWORK	
Steelwork	G10 Structural steel framing
	G11 Structural aluminium framing
	G12 Isolated structural metal members
Protection	A42 Contractor's services and facilties
Q METALWORK	
Composite items curtain walling	H11 Curtain walling
	H12 Plastics glazed vaulting/walling
	H13 Structural glass assemblies
windows	L11 Metal windows/rooflights/screens/louvres
	L12 Plastics windows/rooflights/screens louvres
doors	L21 Metal doors/shutters/hatches

SMM6/SMM7 TABLE

SMM6	SMM7
	L22 Plastics/Rubber doors/Shutters/ Hatches
rooflights	L11 Metal windows/rooflights/screens
	L12 Plastic windows/rooflights/screens/ louvres
balustrades and staircases	L31 Metal stairs/walkways/balustrades
sundries	
duct covers etc.	P31 Holes/Covers/Chases/Supports for services
gates/shutters/hatches	L21 Metal doors/shutters/hatches
cloackroom fittings etc.	N10 General fixtures/furnishings/ equipment
steel lintels	F30 Accessories/Sundry items for brick/ block/stone walling
Plates, bars etc.	
floor plates	L31 Metal stairs/walkways/balustrades
matwells	N10 General fixtures/furnishings/ equipment
Sheet metal, wiremesh and expanded metal	L21 Metal doors/shutters/hatches
Holes, bolts, screws and rivets	E42 Accessories cast into in situ concrete
	G20 Carpentry/Timber framing/First fixings
	L10-L12 Windows
	L20-L22 Doors
	L31 Metal stairs/walkways/balustrades
Protection	A42 Contractor's services and facilities

SMM6	SMM7

R PLUMBING AND MECHANICAL
   ENGINEERING INSTALLATIONS

R DISPOSAL SYSTEMS

S PIPED SUPPLY SYSTEMS

T MECHANICAL HEATING/COOL-
   ING REFRIGERATION SYSTEMS

U VENTILATION/AIR CONDITION-
   ING SYSTEMS

X TRANSPORT SYSTEMS

Y MECHANICAL AND ELECTRICAL
   MEASUREMENT

Classification of work
a. rainwater installation
b. sanitary installation
   (including traps)

R10 Rainwater pipework/gutters
R11 Foul drainage above ground

c. cold water installation

S10 Cold water
S13 Pressurized water
S14 Irrigation
S15 Fountains/Water features
S20 Treated/Deionized/Distilled water
S21 Swimming pool water treatment

d. firefighting installation

S60 Fire hose reels
S61 Dry risers
S62 Wet risers
S63 Sprinklers
S64 Deluge
S65 Fire hydrants
S70 Gas fire fighting
S71 Foam fire fighting

e. heated, hot water
installations etc.

S11 Hot water
S12 Hot and cold water (small scale)
S51 Steam

SMM6	SMM7

R   PLUMBING AND MECHANICAL
ENGINEERING INSTALLATIONS (cont'd)

	T20 Primary heat distribution
	T30 Medium temperature hot water heating
	T31 Low temperature hot water heating
	T32 Low temperature hot water heating (small scale)
	T33 Steam heating
f. fuel oil installation	S40 Petrol/Oil-lubrication
	S41 Fuel oil storage/distribution
	S32 Natural gas
g. fuel gas installation	S33 Liquid petroleum gas
h. refrigeration installation	T61 Primary/secondary cooling distribution
	T70 Local cooling units
	T71 Cold rooms
	T72 Ice pads
j. compressed air installation	S30 Compressed air
	S31 Instrument air
k. hydraulic installation	-
l. chemical installation	-
m. special gas installation	S34 Medical/Laboratory gas
n. medical suction installation	S50 Vacuum
p. pneumatic tube installation	S50 Vacuum
q. vacuum installation	R30 Centralized vacuum cleaning
	S50 Vacuum
r. refuse disposal installation	R14 Laboratory/Industrial waste drainage
	R20 Sewage pumping
	R21 Sewage treatment/sterilization
	R31 Refuse chutes
	R32 Compactors/Macerators

SMM6	SMM7

R PLUMBING AND MECHANICAL ENGINEERING INSTALLATIONS cont'd

R33 Incineration plant

s. air handling installation

T40 Warm air heating
T41 Warm air heating (small scale)
T42 Local heating units
T50 Heat recovery

U10 General supply/extract
U11 Toilet extract
U12 Kitchen extract
U13 Car parking extract
U14 Smoke extract/Smoke control
U15 Safety cabinet/Fume cupboard extract
U16 Fume extract
U17 Anaesthetic gas extract
U20 Dust collection
U30 Low-velocity air conditioning
U31 VAV air conditioning
U32 Dual-duct air conditioning
U33 Multi-zone air conditioning
U40 Induction air conditioning
U41 Fan-coil air conditioning
U42 Terminal re-heat air conditioning
U43 Terminal heat pump air conditioning
U50 Hybrid system air conditioning
U60 Free standing air conditioning units
U61 Windows/Wall air conditioning units
U70 Air curtains

Y24 Trace heating
Y25 Cleaning and chemical treatment

t. automatic control installation

Y53 Control components - mechanical

u. special equipment, eg kitchen equipment

N12 Catering equipment

SMM6	SMM7

R **PLUMBING AND MECHANICAL ENGINEERING INSTALLATIONS** (cont'd)

v. other specialist
  installations

X10 Lifts
X11 Escalators
X12 Moving pavements
X20 Hoists
X21 Cranes
X22 Travelling cradles
X23 Goods distribution/Mechanized
    warehousing
X30 Mechanical document conveying
X31 Pneumatic document conveying
X32 Automatic document filing and
    retrieval

Gutterwork

R10 Rainwater pipework/gutters

Pipework (including fittings)
Ductwork

Y10 Pipelines
Y30 Air ductlines
Y31 Air ductline ancillaries
Y40 Air handling units
Y41 Fans
Y42 Air filtration
Y43 Heating/Cooling coils
Y44 Humidifiers
Y45 Silencers/Acoustic treatment
Y46 Grilles/Diffusers/Louvres

Equipment and ancillaries

N13 Sanitary appliances/fittings

T10 Gas/Oil fired boilers
T11 Coal fired boilers
T12 Electrode/Direct electric boilers
T13 Packaged steam generators
T14 Heat pumps
T15 Solar collectors
T16 Alternative fuel boilers

SMM6	SMM7

R PLUMBING AND MECHANICAL ENGINEERING INSTALLATIONS (cont'd)

	T60 Central refrigeration plant
	Y11 Pipeline ancillaries
	Y20 Pumps
	Y21 Water tanks/cisterns
	Y22 Heat exchangers
	Y23 Storage cylinders/calorifiers
	Y52 Vibration isolation mountings
	Y54 Identification - mechanical
Insulation	Y50 Thermal insulation
Sundries	Y51 Testing and commissioning of mechanical services
	Y59 Sundry common mechanical items
Builder's work	P31 Holes/Chases/Covers/Supports for services
Protection	A42 Contractor's services and facilities
	Y59 Sundry common mechanical items

S ELECTRICAL INSTALLATIONS

V ELECTRICAL SUPPLY/POWER LIGHTING SYSTEMS

W COMMUNICATIONS/SECURITY/ SYSTEMS

Y MECHANICAL AND ELECTRICAL SERVICES

Classification of work	
a. incoming services	V11 HV supply/distribution/public utility supply
	V12 LV supply/public utility supply
b. standby equipment	V32 Uninterrupted power supply
	V40 Emergency lighting

SMM6	SMM7

S   ELECTRICAL INSTALLATIONS (cont'd)

c. mains installation excluding final sub circuits	V20 LV distribution
d. power installation	V22 General LV power
e. lighting installation	V21 General lighting V41 Street/Area/Flood lighting V42 Studio/Auditorium/Arena lighting V90 General lighting and power (small scale) W21 Projection W22 Advertising display
f. electric heating installation	V50 Electric underfloor heating V51 Local electric heating units
g. electric appliances	
h. electrical work associated with plumbing and mechanical engineering installations	Y53 Control components - mechanical
j. telephone installations	W10 Telecommunications
k. clock installation	W23 Clocks
l. sound distribution installation	W11 Staff paging/location W12 Public address/Sound amplification W13 Centralized dictation
m. alarm system installation	W41 Security detection and alarm W50 Fire detection and alarm
n. earthing system installation	W51 Earthing and bonding

SMM6/SMM7 TABLE

SMM6	SMM7
p. lightning protection installation	W52 Lightning protection
q. special services	V30 Extra low voltage supply
	V31 DC supply
	W20 Radio/TV/CCTV
	W30 Data transmission
	W40 Access control
	W53 Electromagnetic screening
	W60 Monitoring
	W61 Central control
	W62 Building automation
Equipment and control gear	V10 Electricity generation plant
	Y70 HV switchgear
	Y71 LV switchgear and distribution board
	Y72 Contactors and starters
	Y92 Motor drives - electric
Fittings and accessories	Y73 Luminaires and lamps
	Y74 Accessories for electric services
Conduit, trunking and cable trays	Y60 Conduit and cable trunking
	Y62 Busbar trunking
Cables/Final sub-circuits	Y61 HV/LV cables and wiring
Earthing	Y80 Earthing and bonding components
Ancillaries	Y82 Identification - electrical
	Y89 Sundry common electrical items
Sundries	Y81 Testing and commissioning of electrical services

SMM6	SMM7

S ELECTRICAL INSTALLATIONS (cont'd)

Builder's work	P31 Holes/Chases/Covers/Supports for services
Protection	A42 Contractor's services and facilities

T FLOOR, WALL AND CEILING FINISHINGS   **M SURFACE FINISHES**

In situ finishings/Lathing and
base boarding/Beds and backings

J10 Specialist waterproof rendering
(including accessories)
J33 In situ glass reinforced plastics

M10 Sand cement/Concrete/Granolithic
screeds/flooring
M12 Trowelled bitumen/resin/rubber-late
flooring

M20 Plastered/Rendered/Roughcast coat
including backings
M21 Insulation with rendered finish
M22 Sprayed mineral fibre coatings
M23 Resin bound mineral coatings
M30 Metal mesh lathing/Anchored
reinforcement for plastered coatings
M41 Terrazzo tiling/In situ terrazzo

Surface sealers   M60 Painting/Clear finishing
Tile, slab and block
finishings/Mosaic work   M40 Stone/Concrete/Quarry/Ceramic
tiling/Mosaic
M41 Terrazzo tiling/In situ terrazzo
M42 Wood block/Composition block/
Parquet flooring
Flexible sheet finishings   M50 Rubber/Plastics/Cork/Lino/Carpet
tiling/sheeting

SMM6/SMM7 TABLE

SMM6	SMM7
Dry linings and partitions	**K LININGS/SHEATHING/ DRY PARTITIONING**
	K10 Plasterboard dry lining
	K30 Demountable partitions
	K31 Plasterboard fixed partitions/inner walls/linings
	K32 Framed panel cubicle partitions
	K33 Concrete/Terrazzo partitions
	K41 Raised access floors
Suspended ceilings, linings and support work	K40 Suspended ceilings
Fibrous plaster	M31 Fibrous plaster
Fitted carpeting	M51 Edge fixed carpeting
Protection	A42 Contractor's services and facilities
U GLAZING	
Glass in openings	L40 General glazing
Leaded lights and copper lights in openings	L41 Lead light glazing
Mirrors	N10 General fixtures/furnishings/ equipment
	N20, N21, N22, N23 Special purpose furnishings/equipment
Patent glazing	H10 Patent glazing
Domelights	L12 Plastics windows/rooflights/screens/ louvres
Protection	A42 Contractor's services and facilities

V PAINTING AND DECORATING

Painting, polishing and similar work	M60 Painting/Clear finishing
Signwriting	N15 Signs/Notices
Decorative paper, sheet plastic or fabric backing and lining	M52 Decorative paper/fabrics
Protection	A42 Contractor's services and facilities

W DRAINAGE	R DISPOSAL SYSTEMS
Pipe trenches	R12 Drainage below ground
Manholes, soakaways, cesspits and septic tanks	R13 Land drainage
Connections to sewers	
Testing drains	
Protection	A42 Contractor's services and facilities

X FENCING	Q40 Fencing
Open type fencing	
Close type fencing	
Gates	
Sundries	
Protection	

EXTERNAL WORKS	D GROUNDWORK
	F MASONRY
	Q PAVING/PLANTING/FENCING/ SITE FURNITURE
	Q10 Stone/Concrete/Brick kerbs/ edgings/channels
	Q20 Hardcore/Granular/Cement bound bases/sub-bases to roads/pavings

SMM6/SMM7 TABLE

SMM6	SMM7
	Q21 In situ concrete roads/pavings/bases
	Q22 Coated macadam/Asphalt roads/pavings
	Q23 Gravel/Hoggin roads/pavings
	Q24 Interlocking brick/block roads/pavings
	Q25 Slab/Brick/Sett/Cobble pavings
	Q26 Special surfacings/pavings for sport
	Q30 Seeding/Turfing
	Q31 Planting
	Q50 Site/Street furniture/equipment

FEDERATION OF MASTER BUILDERS

National President: *E. G. Evans*
Director General : *J. D. Maiden*

Gordon Fisher House
14/15 Great James Street
London WC1N 3DP
Telephone: 071-242 7583
Facsimile : 071-404 0296

Dear Colleague,

Today, more than ever before, we need to have confidence in the method of assessing the basic costs of our work before submitting a quotation to the client. Endeavouring to earn a living in today's economic climate with its low state of enquiries and narrow margins makes it essential to be aware of all cost factors before entering into the field of tough competition.

You will be aware that, for the smaller and medium size companies making up the membership of the *Federation of Master Builders*, the ease and speed of access to these cost factors can save in office time as well as giving the speed and accuracy required by today's prospective client.

As a user of *Spon's Contractors' Handbook Minor Works, Alterations, Repairs and Maintenance*, I can recommend its use to you.

The book will be of benefit to both the newer builder and the well-established contractor. Its sections on the business side of the industry will prove very useful and the Rates for Measured Work will give an excellent basis for the estimator.

I am sure that you will gain a great deal of benefit from this book in the short and long terms and that business will improve in the former.

Yours sincerely,

Edward G. Evans
National President

Introduction

There are two main ways that contractors prepare tenders and quotations. The first and probably the less frequent method is to insert rates against item descriptions in a tender document prepared by the prospective employer's professional advisers. This would occur on major schemes where a bill of quantities and/or a schedule of rates has been prepared. In this case the contractor would be able to examine and use the rates contained in Chapter 5 of this book as a basis for his bid. Usually, however, the contractor is either handed a plan and specification or merely invited to inspect the premises and prepare his offer without the benefit of any paperwork at all! In either of these two cases the contractor must take off his own quantities. Once he has done that, the contents of Chapter 5 can be used as a base on which he can prepare his quotation.

Whichever method is used the main value of this book lies in a sensible application of the thousands of rates in Chapter 5. People who have not used price books before sometimes query their value in that every craftsman does not work at the same speed, wide variations in discounts for materials can be obtained, each job has different cost related circumstances etc.

The answer to these criticisms is that successful users of price books are fully aware of these difficulties but overcome them by understanding the relationship between their own production costs and material discounts and those assumed in the book.

For example, a regular user of this book will know that the published rates may be a certain percentage higher or lower than his own costs. With this knowledge he can quickly prepare his quotations by using the book rates and making the appropriate percentage adjustment at the end.

The editors regularly receive comments from contractors using these books and it seems that most firms' costs fall in a band between 10% higher and lower of those in this book. Some firms have been good enough to state that they have been able to win contracts in competition by using the book's rates verbatim and have secured good profits as well.

INTRODUCTION

Careful thought must be given to the unique circumstances of each job undertaken and the percentage allowances for 'project labour factors' at the beginning of Chapter 5 should be studied carefully and their potential cost in all quotations.

Profits can turn into losses without an assessment of the effect that each project's elements can have on the rates and the editors urge readers to consider this matter very carefully.

Someone once said that price books were like guns - dangerous in the wrong hands! There is some truth in this but using the information wisely can save the one commodity you cannot buy - time.

And that in itself should be of great value to a busy contractor.

Chapter 1
Starting up in Business

Before committing himself by giving up his job, the would-be businessman should consider carefully whether he has the skills and also the temperament to survive in the highly competitive self-employed market. He should also do a lot of research and seek out as much information as possible about how to run a business. He should then know whether his business idea is likely to work in practice and have some idea of the new, strange and sometimes complex requirements of running such a business.

INITIAL RESEARCH

Matters to be researched should include the following.

Finance

Assess what funds will be needed and when will such items as premises, plant, transport, tools, initial stocks of materials, wages, overheads and the proprietors' living expenses need to be paid for before the cash from work done begins to flow in. How is the price of work done to be calculated?

All of this needs a proper 'business plan' and if the bank or other sources of finance are to be tapped such a plan is essential. Fortunately help in its preparation is available from a number of agencies sponsored by goverment, local authorities and industry, usually at little or no cost.

Testing the market

Talk to as many traders already operating in the same field as possible. Try to identify if the need is in the industrial, commercial, local government or domestic field. Talk to likely customers and clients and consider whether it is possible to improve on what they are being offered at present in terms of price, quality, speed, convenience, follow-up, etc.

STARTING UP IN BUSINESS

Advertising

Those entering the domestic side of the business will need to think about the best way to reach potential customers. Are local word-of-mouth recommendations enough to provide reasonably continuous work? If not, what would be the most effective method?

Advertising is costly and it is a waste of funds to place an advert in a paper which circulates in areas A, B, C and D if the business can only cover area A. Advice on the best medium and the content of adverts is available at the Small Business Service.

Experience and training

After making an objective appraisal of the likely market for the goods or services, then the extent to which one is equipped to satisfy its requirements should be considered. Gaps in experience might be filled by a change of present duties or of employer, whilst some lack of skills can probably be overcome by taking a training course.

INITIAL INFORMATION

Training and Enterprise Councils (TECs)

There is no shortage of information about the many aspects of starting and running your own business: finance; marketing; legal requirements; developing your business idea; taxation, etc., are all the subject of a mountain of books, pamphlets, guides and courses. Indeed the likelihood is that the aspiring businessman will be overloaded and thoroughly confused rather than left high and dry without guidance. Nor is it necessary to pay out a lot of money.

A good place to start for both information and advice is your local TEC. This comprises a board of directors drawn from the top men in local industry, commerce, education, trade unions etc., who, together with their staff and experienced business counsellors, assist both new and established concerns in all aspects of running a business. This takes the form not only of across-the-table advice but also, if desired, hands-on assistance in management, marketing, finance etc. There are also training courses and seminars available in most areas.

Contact can be made through the local job centre, Citizens' Advice Bureau or by ringing Freephone Enterprise 0800 222999.

INITIAL INFORMATION

Business links

There are organizations currently being established with a view to providing a 'one-stop-shop' for advice and assistance to owner-managed businesses. When established they will often replace the need to contact TECs and many of the other offical organizations listed below.

Enterprise Agencies

Another useful source of free help and advice is the local Enterprise Agency. Run by local businesses for small and developing concerns it covers similar ground to the TEC.

The address and telephone number of the agency in your area can be obtained by ringing 'Business in the Community' on 071-253 3716.

The Business Start-up Scheme

This is an allowance of £40 a week, in addition to any income made from your business, paid for forty weeks.

To qualify you must be at least eighteen and under sixty-five, work at least thirty-six hours per week in the business and have been unemployed for at least six weeks or fall into one of the other categories - disabled, ex-HMF, redundant etc.

The first step is to get the booklet on the subject from your local job centre or TEC; all the details are in it, including how and where to apply. Once in receipt of the Enterprise Allowance you will also have the benefit of advice and assistance from an experienced businessman from your TEC. All the initial counselling services and training courses are free.

Potential customers and trade contacts

Many who become self-employed in the construction industry already have experience as employees. Use these contacts to check the market, establish the sort of work which is available and the current contract rates. In the domestic market check on the competition for prices, standards of work and service provided, customer complaints and types of advertising.

Try to get firm promises of work before the start-up date.

Potential suppliers

Canvass local suppliers for the best prices, credit terms, minimum order, discounts offered and delivery times. Remember cash is the life blood of business and a supplier who gives 30 days' credit and delivers small quantities at 24/48 hours' notice may be a better buy than one with lower prices but who operates on cash and carry terms only. It is important not to overstock, but carry only what is needed for current requirements. Do not tie up scarce and expensive money in stock which may not be used for weeks or even months.

Banks

Approach banks for information about a business current account and financial services available. Find out about the types of loan required; from an overdraft for working capital to medium and long terms loans including the Government's Loan Guarantee Scheme (see page 10) for the purchase of plant and machinery and alterations to premises. See the TEC first for information about the best approach and find out what the bank manager will need. Shop around several banks and branches if you are not satisfied at first; managers vary widely in their views of what is a viable business proposition. Most banks have useful free information packs to help business start-up.

Point of contact: the local bank manager.

HM Inspector of Taxes

Make a preliminary visit to the local tax office enquiry counter for their publications:

IR 14/15	*Construction Industry Tax Deduction Scheme*
IR28	*Starting in Business*, and, if needed,
IR40	*Conditions for Getting a Sub-contractor's Tax Certificate*
IR53	*PAYE for Employers* (if you employ someone)
IR56/N139	*Employed on Self-employed.*

The onus is on the taxpayer to notify the Inland Revenue that he is in business. Failure to do so may result in the imposition of interest and penalties. Either send a letter or use the form provided in the middle of the *Starting in Business* booklet.

Point of contact: telephone directory for address.

VAT Office

Registration for VAT is required if:

1. At the end of any month the value of taxable supplies in the past twelve months has exceeded the annual threshold;

4

2. There are reasonable grounds for believing that the value of taxable supplies in the next 30 days will exceed the annual threshold.

Taxable supplies include any zero-rated items. From 1 December 1993 the annual threshold is £45,000. Failure to register is an offence punishable by the imposition of financial penalties.

The VAT office also carries a number of useful publications including:

700	*The VAT Guide*
700/1	*Should I be Registered for VAT?*
700/12	*Filling in Your VAT Return*
700/21	*Keeping Records and Accounts*
708/2	*Application of VAT to the Construction Industry*
731	*Cash Accounting*
732	*Annual Accounting*
742	*Land and Property*

Notes on the application of VAT to the Construction Industry are in Chapter 4 together with information about the new 'Cash accounting scheme' and the introduction of annual VAT returns.

Point of contact: telephone directory for address.

DSS Office

Class 2 contributions payable by the self-employed may be paid either in cash or by direct debit through a bank account. Call at the local office to make the necessary arrangements. Class 4 contributions which are also payable by the self-employed are collected along with the income tax by the Inland Revenue and no special action by the businessman is required.

Ask at the DSS office for the following publications:

NI 41	*NI Guide for the Self-employed*
NI 27A	*People with Small Earnings from Self-employment*
NI 35	*NI for Company Directors*
NI 225	*Direct Debit - The Easy Way to Pay*, and for employers
NP 15	*Employers' Guide to NIC*
NI 227	*Employers' Guide to Statutory Sick Pay*

Point of contact: telephone directory for address.

START UP IN BUSINESS

Local authorities

Authorities vary in the provisions made for small businesses but all have been asked to simplify and cut delays in planning applications. In Assisted Areas and Enterprise Zones rent-free periods and reductions in rates may be available on certain industrial and commercial properties. As a preliminary to either purchasing or renting business premises the following booklets will be very helpful. *A Step by Step Guide to Planning Permission for Small Businesses* and *Business Leases and Security of Tenure* are both issued by the Department of Employment and are available at council offices, Citizens' Advice Bureaux and TEC offices.

Some authorities run training schemes in conjunction with local industry and educational establishments.

Point of contact: usually the planning department - ask for the Industrial Development or Economic Development Officer.

Department of Trade and Industry

The current package of assistance from this department is called the Enterprise Initiative and is geared more to the needs of existing businesses than new start-ups. It ranges widely over all aspects; marketing, design, quality, management, finance, etc., and consists essentially of assessing the requirements of the business. The counsellor will keep an eye out for untapped resources, inefficient work systems and unrealized potential and will recommend specialist consultants to come in and advise. The department would pay all costs of the initial survey and one half (or two thirds in development areas) of the costs of any consultancy between 5 and 15 man days.

Point of contact: phone 071-215 5000 and ask for the address and phone number of the nearest DTI office and copies of their explanatory booklets.

Department of the Environment

From 1 April 1992 new regulations are in force under the Environmental Protection Act 1990 relating to all forms of waste other than normal household rubbish. Any concern which produces, stores, treats, processes, transports, recycles or disposes of such waste has a 'duty of care' to ensure it is properly discarded and dealt with. Practical guidance on how to comply with the law (it is a criminal offence punishable by a fine not to) is contained in a booklet *Waste Management: The Duty of Care: A Code of Practice*, obtainable from HMSO Publications Centre, PO Box 276, London SW8 5DT. Phone 071-873 9090. Price £5.00.

INITIAL INFORMATION

Accountant

The services of an accountant are to be strongly recommended from the beginning because the legal and taxation requirements start immediately and must be properly complied with if trouble is to be avoided later. A qualified accountant must be used if a limited company is being formed and for all types of business the accountant should be able to give advice on a whole range of business issues from, for example, book-keeping to grant aid, from tax planning and compliance to finance raising and will clearly help in preparing annual accounts.

It is worth spending some time finding an accountant who has other clients in the same line of business and is able to give sound advice particularly on taxation and business finance and is not so overworked that damaging delays in producing accounts are likely to arise. Ask other traders whether they can recommend their own accountant. Visit more than one firm of accountants, ask about the fees they charge and how much the production of annual accounts and agreement with the Inland Revenue are likely to cost and how long the work will take. A good accountant is worth every penny of his fees but do not hesitate to challenge him if his service is unsatisfactory.

Solicitor

Many businesses operate without the services of a solicitor but there are a number of occasions when legal advice should be sought. In particular no-one should sign the lease of premises without asking a solicitor what they are committing themselves to because it is not unusual for a business to be put into financial difficulty through unnoticed liabilities in its lease. Either an accountant or solicitor will help with drawing up a partnership agreement which all partnerships should have.

A solicitor will also help to explain complex contract terms and prepare draft contracts if the type of business being entered into requires them.

Insurance broker

Policies are available to cover many aspects of business including:

employer's liability - compulsory if the business has employees;
public liability - essential in the construction industry;
motor vehicles;
theft of stock, plant, money, etc.;
fire and storm damage;
personal accident and loss of profits.

Brokers are independent advisers who will obtain competitive quotations on your behalf. See more than one broker before making a decision - their advice is normally given free and without obligation.

Point of contact: telephone directory or write for a list of local members to:

> The British Insurance Brokers Association
> Consumer Relations Department
> BIBA House
> 14 Bevis Marks
> London EC3A 7NT (phone: 071-623 9043)

or contact

> The Association of British Insurers
> 51 Gresham Street
> London EC2V 7HQ (phone: 071-600 3333)

who will supply free, a package of very useful advice files specially designed for the small business.

The Health and Safety Executive

The Executive operates the legislation covering everyone engaged in work activities and the following free literature is available:

HSE 16   *The Law on Health and Safety at Work*
IND(G)14(L) *Compliance with Health and Safety Legislation at Work*
HSE4   *Short Guide to Employer's Liability (Compulsory Insurance) Act 1969*

The Executive has issued a very useful set of 'Construction Health Hazard Information Sheets' covering such topics as handling cement, lead and solvents, safety in the use of ladders, scaffolding, hoists, cranes, flammable liquids, asbestos, roofs and compressed gases, etc. A pack of these may be obtained free from your local HSE office or:

> The Health & Safety Executive
> St. Hugh's House
> Stanley Precinct
> Bootle
> Merseyside L20 3QY (phone: 051-951 4381)

FINANCE

Working out how much will be needed

The businessman should estimate in advance what funds will be needed to start up and to run the business for at least the first 12 months. If the forecast is to be reasonably accurate he must make some early decisions about:

1. The premises where the business will be based, what initial repairs, alterations, etc., to them are required and what will be the total cost;

2. What plant, equipment and transport is needed, whether it is to be leased or purchased and again what the cost will be;

3. How much stock of materials, if any, should be carried (the bare minimum only should be acquired so a reliable supplier should be sought out);

4. What will be the weekly bill for overheads, wages and the proprietor's living costs and how these are to be met until the cash for work done starts to flow in;

5. What type of work is going to be undertaken, how much profit margin can realistically be obtained and how often invoices are to be presented.

The above are just some of the items that should be covered. If the proprietor is fortunate enough to have some capital, or some is available within the family, he might get by without doing more than a rough estimate based on his initial research and knowledge of the trade. If however there is the need to seek finance from a bank or other financial institution a much more detailed statement will be required for which the help of an accountant or TEC adviser should be sought.

Sources of funds

Finance, like charity, often begins at home and the would-be businessman should make a realistic assessment of his net worth including the value of his house after deducting any mortgage(s) outstanding on it, his savings, any car or van owned and any sums which his family are prepared to contribute and deduct any private borrowing which will come due for payment in the next 24 months. The whole of these funds may not of course be available (for instance, money which has been loaned to a friend or relative who is known to be unable to repay at the present time). It may not be desirable that it should all be put at risk on a business venture. Establish therefore (a) how much cash it is proposed to invest in the business and (b) whether the family home will be made available for any business borrowings.

Whilst it may be wise not to pledge too much of the family assets, it has to be remembered that the bank will be looking closely at the degree to which the proprietor has committed himself to the venture and will not be impressed by an applicant for a loan who is prepared to risk only a small fraction of his own resources.

Having decided how much of his own funds to contribute, the businessman can now see the level of shortfall and consider how best to fill it. Consideration should be given to partners where the shortfall is large and particularly when there is a need for heavy investment in fixed assets such as premises and capital equipment. It may be worthwhile starting a limited company with others also subscribing capital and to allow the banks to take security against the book debts. The first outside source of money to which most businessmen turn is the bank and a book could be written solely on the do's and don'ts of approaching a bank manager. Here are a few tips:

1. Have a proper business plan to present to him including a cash flow forecast (12 months is usual), also an opening statement of affairs, projected profit and loss accounts and balance sheets for two years and a written statement describing the whole business venture. Use conservative estimates which tend to understate rather than overstate the forecast sales and profits.

2. Know the figures in detail - and don't leave it to an accountant to explain them for you. The bank manager is interested in the businessman not his advisers and will be impressed if he has a sound grasp of the financing of his business.

3. Understand the difference between short and long term borrowing; know how much is needed of each and be ready to explain how the business will be able to repay the bank its money.

4. Ask about the Government Loan Guarantee Scheme if there is a shortage of security for loans. Under this scheme the Government guarantees 70% of the loan up to £100,000 (now up to 85% for loans to businesses in the Inner City Task Force Areas) but if you have been in business for more than two years the limit is £250,000. In return there is an insurance of 0.5% on fixed rate interest arrangements and 1.5% on variable bank interest arrangements. Repayment is over a period from 2 to 7 years.

There are a number of other financial institutions in the 'venture capital' market that can help well established businesses, usually limited companies, wishing to expand and also for some well conceived start-ups. They will provide a flexible package of equity and loan capital in the range of £25,000 to £1,000,000. Usually the deal entails the institution having a minority interest in the voting share capital and a seat on the board of the company. Arrangements for the eventual purchase of the shares held by the finance company by the private shareholders are also normally incorporated in the scheme.

Contact points for information and advice are bank managers, accountants and the TEC. If the outside investor in the business is an individual he will probably wish to invest within the terms of the Enterprise Investment Scheme which enables him to get tax relief at 20% on the amount of his investment. The rules are complex and professional advice is essential.

The Royal Jubilee and Prince's Trusts

These trusts, through the Youth Business Initiative, provide bursaries of not more than £1,000 per individual to selected applicants who are unemployed and aged 25 or under. Grants may be used for tools and equipment, transport, fees, insurance, instruction and training, but not for working capital, rent and rates, new materials or stock. They operate through a local representative whose name and address may be ascertained by contacting The Prince's Youth Business Trust, 5 Cleveland Place, London. SW1Y 6JS: Phone 081-968 3713.

THE CONSTRUCTION INDUSTRY TAX DEDUCTION SCHEME

General

The Construction Industry Tax Deduction Scheme is known universally as the '714' Scheme, after the number of the official form around which the whole system revolves. The government is proposing important changes to the scheme but they are still under discussion.

The businessman should visit his local income tax enquiry office and obtain copies of the Revenue booklet IR 14/15 and leaflet IR 40 which explain the conditions under which the Revenue will issue a 714 certificate.

The scheme operates whenever a 'contractor' makes a payment to a 'sub-contractor'.

If the sub-contractor does not hold a valid tax certificate (714I, 714P, 714C or 714S) issued to him by the Inland Revenue then the contractor *must* deduct 25% tax from the whole of any payment made to him (excluding the cost of any materials). If, however, he holds such a certificate the payment may be made in full without deducting tax (in the case of a 714S there is a weekly limit after which tax is deductible).

A business is not obliged by law to seek an exemption certificate and can legally work in the construction industry without one. As a matter of practice, however, many main contractors are reluctant to undertake the additional paperwork required when tax has to be deducted and accounted for to the Revenue and will give work to a sub-contractor with a 714 certificate in preference to one without.

A small business that does work only for the general public and small commercial concerns, is outside the scheme and does not need a 714 certificate to trade. If, however, it engages other contractors to do jobs for it, the business would have to register under the scheme as a contractor and deduct tax from any payments made to a sub-contractor who did not produce a valid 714 certificate.

Obtaining a 714 certificate

There is a special application form which may be obtained from tax offices. There are a number of conditions which have to be met before the Revenue will issue a certificate. In general terms these are:

1. The applicant must be working as a sub-contractor in the UK in the construction industry;

2. The business must be run from proper premises with the usual business facilities, and records must be kept and a business bank account operated;

3. The applicant must have been employed or self-employed in the UK for a continuous period of 3 years in the 6 years before the date of application; short breaks in employment will not be taken into account unless they exceed 6 months in total (but see 5);

4. The applicant must have a satisfactory tax and NIC record: (the Inland Revenue will check the NI contribution position with the DSS);

5. School leavers who can show that in the 6 years up to the date of application they were in full-time education or training for a continuous period of 3 years, or in full-time education or training for part of the 3 years and unemployed or in self-employment for the rest, may apply for a special certificate (714S);

6. Those who satisfy all the conditions except 3 may also obtain a special certificate (714S) if they can arrange for a bank to guarantee the tax payable on the amounts received in full (leaflet IR 40 sets out the conditions in details).

The 714 certificate

There are four types:

714I - which is issued to individuals;
714P - which is issued to partners;
714C - which is issued to most limited companies;
714S - which is issued as explained above.

The I, P and S certificates include a photograph of the individual to whom it is issued in addition to his name, NIC number, signature and business name (if any).
The certificate also has a serial number and an expiry date. Vouchers numbered 715 (in the case of a special certificate, 715S) are issued along with the 714I, P and S.

The 714 system in operation

Holders of certificates 714I and P (including company directors holding 714Ps).

Before paying a sub-contractor in full, without any deduction of tax, the contractor must carry out detailed checks on the 714 certificate (the original, not a copy) and the sub-contractor must produce the 714 for this purpose. On being paid the sub-contractor must complete a voucher 715 and give it to the contractor, showing, among other things, the amount of the payment received. The contractor must send the 715s to the Revenue monthly unless his monthly deduction plus any PAYE and NIC for employees total less than £450, when he may choose to pay quarterly.

Holders of 714C certificates

A sub-contractor which is a limited company may choose to produce to the contractor either the 714C itself or a 'certifying document'. Whichever method is used the documents have to be checked in detail by the contractor, if necessary by telephoning the company itself to confirm that the person presenting the document is their authorized representative.
If the contractor is satisfied he must pay the sub-contractor in full; 715s

are not used but both parties record the details of the payment in their business records.

Holders of 714S certificates

If the weekly payment to the sub-contractor (excluding the cost to him of any materials purchased directly from another person) does not exceed the limit shown on the front of the 715S voucher, then the procedure outlined above for a certificate 714I and P is followed and a voucher 715S is given to the contractor on receipt of the payment.

If, however, the payment exceeds the amount shown for the week then the 714I and P procedure is followed in respect of the amount shown and the procedure below is carried out for the balance of the payment.

Sub-contractors with no certificate and those with S certificate payments in excess of the weekly limit

The contractor is obliged to deduct tax from all payments (excluding the cost of directly purchased materials) and to account to the Revenue for all amounts so withheld. To enable the sub-contractor to prove to the inspector of taxes that he has suffered this tax deduction the contractor must give him a certificate on form SC60 showing the amount withheld. These SC60 forms must be carefully filed for production to the Inspector after the end of his accounting year along with his business profit and loss account and balance sheet. Any tax deducted in this way over and above the sub-contractor's proper, agreed liability for the year will be repaid by the Inland Revenue.

At the end of the day the sub-contractor will pay the same amount of income tax whether or not he has a certificate, but those without one will have suffered a severe restriction in their cash flow until the repayment is made.

The main contractor periodically has to send the 715 vouchers to the Inland Revenue Computer Centre in Liverpool and make an annual return to that Centre also. The date for filing the end-of-year return of payments to uncertificated sub-contractors has been extended by one month to 19 May. It all amounts to a heavy burden on the trader with penalties awaiting those who are sloppy, dilatory or dishonest in its operation.

Miscellaneous points

1. A payment includes anything paid out by the contractor such as a 'sub' or a loan, whether by cash, cheque or credit and whether direct to the sub-contractor or to his nominee.

14

2. The cost of subsistence and travelling expenses reimbursed by the contractor is included in the amount on which the tax deduction is calculated.

3. The scheme is policed by the Revenue in much the same way as they inspect PAYE documents, and records for the scheme have to be made available on request.

4. A contractor is liable to pay to the collector of taxes all amounts which he *should* have deducted from sub-contractors, whether he made the deductions or not. He may, however, be excused from having to pay if he can show he took reasonable care and made the error in good faith.

   If the deductions have not been properly made the sub-contractor himself will be asked to account for his own correct liability.

5. In a situation where tax *has* been deducted from payments made by him but the contractor has failed to account for it to the collector, the sub-contractor would be unlikely to recover any over-deductions from the Revenue.

6. Disputes between contractors and sub-contractors about the amount of any deductions should be referred to the inspector for a ruling.

Note
   The above is merely a summary of the very detailed instructions contained in IR 14/15 paragraphs 65-129 which must be carefully studied by anyone involved in operating the scheme either as a contractor or sub-contractor.

VAT

The general rule about liability to register for VAT is given in the VAT office notes above. It is possible to give here only a brief outline of how the tax works. The rules which apply to the construction industry are extremely complex and all traders must study *The VAT Guide* and other publications.

   The amount of tax to be paid is the difference between the VAT charged out to customers *(output tax)* and that suffered on payments made to suppliers for goods and services *(input tax)*. Unlike income tax there is no distinction in VAT for capital items so that the tax charged on the purchase of, for example, machinery, trucks and office furniture will normally be reclaimable as Input Tax. One important exception to this is that the input tax on a car cannot be reclaimed even though it is used wholly for business purposes.

   VAT is payable in respect of 3-monthly periods known as 'tax periods' and you can apply to have the group of tax periods which fits in best with your financial year. The tax must be paid within one month of the end of each tax period. Traders who receive regular repayments of VAT can apply to have them monthly rather than quarterly.

Not all types of goods and services are taxed at 17.5% (i.e., at the standard rate), some are exempt and others are zero-rated.

Zero-rated

This means that no VAT is chargeable on the goods or services but a registered trader can reclaim any *input* tax suffered on his purchases. For instance a builder pays VAT on the materials he buys, but if he is constructing a new dwelling house, this is zero-rated and he may reclaim this VAT or set it off against any VAT due on other standard rated work.

Exempt

Supplies which are exempt are less favourably treated than those which are zero-rated. Again no VAT is chargeable on the goods or services but the trader cannot reclaim any *input* tax suffered on his purchases.

Standard-rated

All work which is not specifically stated to be zero-rated or exempt is standard-rated, i.e., VAT is chargeable at the current rate of 17½% and the trader may deduct any *input* tax suffered when he is making his return to the Customs and Excise.

If for any reason a trader makes a supply and fails to charge VAT when he should have done so (e.g. mistakenly assuming the supply to be zero-rated) he will have to account for the VAT himself out of the proceeds. If there is any doubt about the VAT position it is safer to assume the supply is standard rated, charge the appropriate amount of VAT on the invoice and argue about it later.

Time of supply

The *time* at which a supply of goods or services is treated as taking place is important and is called the 'tax point'. VAT must be accounted for to the Customs and Excise at the end of the accounting period in which this 'tax point' occurs. For the supply of *goods* which are 'built on site' the 'basic tax point' is the date the goods are made available for the customer's use, whilst for *services* it is normally the date when all the work except invoicing is completed.

VAT

However, if you issue a tax invoice or receive a payment *before* this 'basic tax point' then that date becomes the tax point.

In the case of contracts providing for stage and retention payments the tax point is either the date the tax invoice is issued or when payment is received, whichever is the earlier.

All the above requirements apply to sub-contractors and main contractors and it should be noted that, when a contractor deducts income tax from a payment to a sub-contractor (because he has no valid 714), VAT is payable on the full gross amount *before* taking off the Income Tax (see Chapter 3).

See below for examples of VAT payments and repayments.

Examples of how VAT works are as follows:

(a) Sales invoices total	£1000
plus VAT @ 17½%	175
The customer pays	£1175
Purchase invoices total	£600
plus VAT @ 17½%	105
Total paid	£705

In this example the *output* tax is £175 and the *input* tax is £105 which means that the trader owes £70 to the Customs and Excise.

If in the tax period machinery costing £2000 plus VAT £105 had also been purchased the figures would be:

(b) Sales invoices total	£1000
plus VAT @ 17½%	175
The customer pays	£1175

17

Purchase invoices total	£2600
plus VAT @ 17½%	<u>455</u>
Total paid	<u>£3055</u>

In this example the *output* tax is also £175 but the *input* tax is £455 so that a refund of £280 is due to the trader from the Customs and Excise.

Chapter 2
Running the business

Many businesses are run without adequate information being available to check trends in their vital areas, e.g. marketing, money and managerial efficiency. It is vital look critically at all aspects of the business to maximise profits and eliminate inefficiency.

Proprietors often have the feeling that the business should be 'doing better' than it is, without being able to identify what is going wrong. Sometimes there is the worrying phenomenon of a steadily increasing work programme coupled with a persistently reducing bank balance or rising overdraft.

Some useful ways of checking the position and identifying problem areas are given below.

Marketing

Whilst management and finance are concerned with the internal running of a business the market is where it makes contact with the outside world in the shape of its competitors and customers. Throughout his business life the entrepreneur should study carefully the methods and approach of the former and the needs and wishes of the latter. A shortcoming frequently found in ailing concerns is that the proprietor thinks he knows better than his customers what they want.

The term 'market research' sounds both difficult and expensive but a very simple form of it can be done quite effectively by the businessman and his sales staff. First, identify the type of person or business to whom the products or service are likely to appeal, finding out and recording what it is the customer wants in terms of price, quality, design, payment terms, follow-up service, guarantees, services.

The initial approach might be by leaflet or letter followed by a personal call. As an on-going part of management all staff with customer contact should be encouraged to enquire about and record customer preferences, complaints, etc., and feed it back to management.

Other sources of information are friends in the trade, business journals, trade exhibitions, suppliers, representatives, etc., from whom information about trends, new techniques and products can be obtained and studied.

Valuable information can also be gained from studying competitors and the following questions should be asked:

- what do they sell and at what prices?

- what inducements to buy do they offer their customers (e.g. credit facilities, guarantees, free offers, discounts, etc.)?

- how do they reach their customers (local/national advertising; mail shots; salesmen; local radio and TV)?

- what are the strongest aspects of their appeal to customers and have they any weaknesses?

The businessman should apply all the information gathered from customers and competitors to his own range of products with a view to making sure he is offering the right product at the right price in the most attractive way and in the most receptive market.

In a small business where the proprietor is also his own salesman he must give careful thought to how he can best present his product and himself. If he is working solely within the construction industry his main problems are likely to centre on getting a 714 and on exploiting as fully as possible trade contacts to get sub-contract work.

However, for those who serve the general public, presentation can be a vital element in getting work. The customer is looking for efficiency, reliability and honesty in a trader and quality, price and style in the product. To bring out these facets in discussion with a potential customer is a skilled task and for newcomers to business and those who have not had as much success to date as they hoped for, a short course on marketing techniques could pay handsome dividends.

The Enterprise unit of the Manpower Services Commission will give the names and addresses of such courses locally - contact it through the job centre or make an appointment to see a TEC adviser with marketing experience and talk over the problems with him.

Some indications that a market review is needed are:

- declining sales

- profit margins being squeezed

- fewer customers and over-reliance on a few large ones

- profitability not spread evenly over the product or service range - some items 'not earning their keep' - perhaps through being out of date, too expensive, badly designed, poorly constructed, etc.

Unfortunately some firms which close down do not seek financial advice until it is too late to halt the downward trend when earlier attention to the problems may have saved some of them. There are many reasons for this and one of them is that those running the business are unable to recognise the tell-tale signs and very few accountants take the trouble to explain to their clients what to look for. There are some tests and checks that can be done quite easily.

Cashflow

Cashflow is the lifeblood of the business and more businesses fail through lack of cash than for any other reason. Cash is generated through the conversion of work into debtors and then into payment and throught the deferral of the payment of supplies for as long a period as can be negotiated.

The objective must be to keep stock, work in progress and debts to a minimum and creditors to a maximum. Trends in important ratios as well as absolute values can help in assessing the business performance.

Debtor days

This is calculated by dividing your trade debtors by annual sales and multiplying by 365. It shows the number of days credit being afforded to your customers and should be compared both with your normal trade terms and the previous month's figures. Normal procedures should involve the preparation of a monthly aged list of debtors showing the name of the customer, the value and which month it relates to.

The oldest and largest debtors can be seen at a glance for immediate consideration of what further recovery action is needed. The list may also show over-reliance on one or two large customers or the need to stop supplying a particularly bad payer until his arrears have been reduced to an acceptable level. Consideration should be given to making up bills to a date before the end of the month and making sure the accounts are sent out immediately followed by a statement 4 weeks later. Consider giving discounts for prompt payment.

If all else fails, and legal action for recovery is being contemplated, call at the County Court and ask for their leaflets numbers 1 to 4.

RUNNING THE BUSINESS

Stockturn

The level of stock should be kept to a minimum and the number of days stock can be calculated by dividing the stock by the annual purchases and multiplying by 365. A worsening trend on a month by month basis shows the need for action. It is important to make regularly a full inventory of all stock and dispose of old or surplus items for cash. A stock control procedure to avoid stock losses and to keep stock to a minimum should be implemented.

Profitability

Whilst cash is vital in the short-term, profitability is vital in the medium-term. The two key percentage figures are the gross profit percentage and the net profit percentage. Gross profit is calculated by deducting the cost of materials and direct labour from the sales figure whilst net profit is arrived at after deducting all overheads.

Possible reasons for changes in the gross profit percentage are:

- the pricing of jobs is not taking full account of increases in materials and wages and a review of pricing policy is needed

- too generous discount terms are being offered

- poor management, overmanning, waste and pilferage of materials

- too much down-time on plant which is in need of replacement.

If net profit is deteriorating after the deduction of an appropriate reward for your own efforts, including an amount for your own personal tax liability, you should review each item of overhead expenditure in detail asking amongst others the following questions:

- can savings be made in non-productive staff?

- is sub-contracting possible and would it be cheaper?

- have all possible energy-saving methods been fully explored?

- do the company's vehicles spend too much time in the yard; can they be shared and their number reduced?

- is the expenditure on advertising producing sales (review in association with 'marketing' above)?

Over trading

Many inexperienced businessmen imagine that profitability equals money in the bank. In some cases, particularly where the receipts are wholly in cash, this may be the case, but additional business may mean higher stock inventories, extra wages, overheads, increased capital expenditure on premises and plant.

If the debtors show a marked increase as the turnover rises the proprietor may find to his surprise that each expansion of trade reduces rather than increases his cash resources. The business, which had enough funds for start-up, finds it does not have sufficient cash to run at the higher level of operation and the bank manager may be getting anxious about the increasing overdraft.

It is essential for those who run a business which operates on credit terms to be aware that profitability does not necessarily mean increased cash availability. Regular monthly management information on marketing and finance as described in this chapter will enable 'over trading' to be recognized and remedial action to be taken early.

If the situation is appreciated only when the bank and other creditors are pressing for money, radical solutions may be necessary such as bringing in new finance, sale and leaseback of premises, a fundamental change in the terms of trade, or even selling out to a buyer with more resources. Professional help from the firm's accountant will be needed, whilst the TEC has counsellors experienced in advising on both the marketing and financial aspects of such situations.

Break-even point

The costs of a business may be divided into two types: variable and fixed.

Variable costs are those which increase or decrease as the volume of work goes up or down and include such items as materials used, direct labour, power, machine tools, etc.

Fixed costs are not related to turnover and are sometimes called 'fixed overheads'. They include rent, rates, insurance, heat and light, office salaries, plant depreciation, etc.; these costs are still incurred even though few or no sales are being made.

RUNNING THE BUSINESS

Many small businessmen run their enterprises from home using family labour as back-up; they sell mainly their own labour and buy materials and hire plant only as required.

By these means they reduce their fixed costs to a minimum and start making profits almost immediately. However, larger firms which have business premises, perhaps a small workshop, an office, vehicles, etc., need to know how much they have to sell to cover their costs and become profitable.

In the case of a new business it is necessary to estimate the figures but where annual accounts are available a break-even chart based on them can be readily prepared.

Suppose the real or estimated figures (expressed in £000s) are:

	%	£000
Sales	100	400
Variable costs	66	265
Gross profit	34	135
Fixed costs	13	50
Nett profit	21	85

$$\text{Break-even point} = \text{Fixed Costs} \div 1 - \frac{\text{Variable Costs}}{\text{Sales}}$$

$$= 50 \div 1 - \frac{265}{400}$$

$$= 50 \div (1 - 0.6625)$$

$$= 50 \div 0.3375$$

$$= £148 \text{ (thousand)}$$

In practice things are never quite as clear cut as the figures above show, but nevertheless this is a very useful tool for assessing not only the break-even point but also the approximate amount of loss or profit arising at differing levels of turnover and also for considering pricing policy.

Chapter 3
Taxation

A new basis of charging tax on business profits took effect from 6 April 1994 for new businesses and from 6 April 1996 for businesses already established at 6 April 1994.

For new businesses there may still be an advantage in having an accounting date early in the tax year, that is shortly after 6 April. The advantage, however, is less than it would have been under the old rules and professional advice will be needed to choose the best accounting date for tax purposes.

For existing businesses at 6 April 1994 the way income tax is charged on business profits is to change radically. For the 1997/98 tax year onwards profits assessed for tax in a given tax year will be the profits of the accounting period ending in that year and not as now those of the preceding year.

Transitional rules will apply for the tax year 1996/97 whereby the profits of two accounting periods will be used to give an average. For example, a self- employed business with 30 April year end, the average period began on 1 May 1994.

Because of the averaging of profits charged to tax, a tax planning opportunity will arise subject to anti-avoidance provisions introduced by the Inland Revenue to penalize the artificial movement of profit into the averaging period.

In particular the following areas may offer tax planning opportunities:

1. the purchase as opposed to the leasing of capital equipment to be used in the business

2. the introduction of a partner into the business

3. where a partnership is considering for sound business reasons, re-financing by way of personal loans rather than a partnership loan.

Professional advice will be needed to take the maximum advantage of these opportunities.

The timing of a cessation of business will still be important particularly until the new system is fully in place. Professional advice should be sought on the timing and tax cost of ceasing business.

Together with changes in the way taxable profits will be measured, the Inland Revenue are introducing a new system of charging and collecting personal tax. Again from the tax year 1996/97 the burden of assessing tax will shift from the Inland Revenue to the individual tax payer. The main features of this new system are as follows:

1.   the onus is on the taxpayer to provide and complete information

2.   the taxpayer will have a choice: he can calculate and pay his tax liability, at the same time as making his return and this will need to be done by the 31 January following the end of the tax year. Alternatively he can send in his tax return much earlier and the Inland Revenue will calculate the tax to be paid on 31 January

3.   the important aspect to the new system is that if the return is late, or the tax is paid late, there will be automatic penalties, imposed on the taxpayer.

Spouses in business

If spouses work in a business, perhaps answering the phone, making appointments, writing business letters, making up bills and keeping the books, they should be properly remunerated for it. Being a payment to a family member the inspector of taxes will be understandably cautious in allowing it in full as a business expense. The payment should be:

1.   actually paid to them, preferably weekly or monthly and in addition to any housekeeping monies

2.   recorded in the business book

3.   reasonable in amount in line with their duties and the time spent on them.

If the wages paid to them exceed £56.99 p.w. Class 1 employer and employee NIC becomes due and if they exceed £3,445 p.a. (assuming they have no other income) PAYE tax will also be payable.

It should also be noted that once small businesses are well established and the spouse's earnings are approaching the above limits, consideration may be given to bringing them in as a partner. This has a number of effects:

1. there is no longer a need to relate the spouse's income (which is now a share of the profits) to the work they do (if any)

2. they will pay Class 2 and Class 4 NIC instead of the more costly Class 1 contributions and PAYE will no longer apply to their earnings

3. but remember - as partners any assets they own are vulnerable to proceedings by partnership creditors.

Premises

Many small businessmen cannot afford to rent or buy commercial premises and run their enterprises from home using part of it as an office where the books and vouchers, clients records, trade manuals, etc. are kept and estimates and plans are drawn up. In these circumstances a portion of the outgoings on the property may be claimed as a business expense.

Car expenses are usually split on a fractional mileage basis between business journeys, which are allowable, and private ones, which are not, and a record of each should be kept. If the business does work only on one or two sites or for only one main contractor the inspector may argue that the true base of operations is the work site not the residence and seek to disallow the cost of travel between home and work. It is tax-wise therefore, and sound business practice, to have as many customers as possible and not work for just one client.

Appeals against assessments. To businessmen, income tax assessments are an anathema. They should resist the temptation to tear them up or put them behind the clock and forget about them. All assessments should be checked for accuracy immediately and, if excessive, the instructions on the notice about making an appeal should be followed and the appeal sent to the Tax District that issued the assessment. The appeal should also show how much of the tax charged should be postponed because the assessment is too high.

If this is not done within 30 days of the issue of the notice, the assessment becomes final and the inspector (and the general commissioners if they are asked to adjudicate) may well not accept a late appeal. In this event the taxpayer has no alternative but to pay the tax as charged on the assessment which may be estimated or contain additions to the profits which have not been agreed by him and his accountant. If the appeal is to be made by the accountant check with him before the 30 days are up to ensure that it has been submitted.

Keep copies of all correspondence with the inspector and collector. Letters can be mislaid or fail to be delivered and it is essential to have both proof of what was sent as well as a permanent record of all correspondence.

TAXATION

Some useful statistics on Income Tax 1994/95

The current personal allowance for a single person is £3,445. The personal allowances for people aged 65-74 and over 75 years are £4,200 and £4,370 respectively. The married couples allowance is £1,720 and £2,665 for a couple between the ages 65 to 74 and £2,705 for a couple over 75 years.

Rates of tax
The value of the married couples allowance has been reduced for 1994/95 as it is now only worth a maximum 20% instead of saving tax at a person's top rate. The rates of tax for 1994/95 are as follows:

> Lower rate: 20% on taxable income up to £3,000
> Basic rate: 25% on taxable income between £3,000 and £23,700
> Higher rate: 40% on taxable income over £23,700

Example: Married man: wife has no income. On earnings of £20,000 he would pay:

		£
Income		20,000
Less personal allowance		3,445
Taxable income		16,555
	£3,000 at 20% =	600
	£13,555 at 25% =	3,389
		3,989
Less married couple's allowance	£1,720 at 20%	344
Tax payable		3,645

On earnings of £30,000 the tax he would pay:

			£
	£3,000 at 20%	=	600
	£20,700 at 25%	=	5,175
	£2,855 at 40%	=	1,142
Less married couple's allowance	£1,720 at 20%		344
	Tax payable	=	6,573

TAXATION

Taxation of husband and wife

A married woman is treated in much the same way as a single person with her own personal allowance and basic rate band. Husband and wife each make a separate return of their own income and the Inland Revenue deals with each one in complete privacy - letters about the husband's affairs will be addressed only to him and about the wife's only to her (unless the parties indicate differently). The allowances and relief are dealt with as follows.

Personal allowance

Husband and wife each has one of these - for 1994/95 it is £3,445 each.

Married couple's allowance

The amount is £1,720. This is initially due to the husband but if his income is too small to use it all he may transfer the surplus part to his wife but once he has passed some allowances in this way he cannot change his mind and ask for them back. Married couples may choose, however, how they wish the allowance to be allocated between them, subject to the right of the wife to claim half the allowance if she wishes. Here are two examples:

Husband and wife both in employment; both earn £7,000 p.a.

1993/94: husband decides to keep all the married couple's allowance. The relief would be:

	Husband £	Wife £
Personal allowance	3,445	3,445
Married couple's allowance	1,720	-
	£5,165	£3,445

1994/95: wife decides she wants her half share of the allowance. The relief would be:

	Husband. £	Wife. £
Personal Allowance	3,445	3,445
Married couple's allowance	860	860
	£4,305	£4,305

31

TAXATION

Basic rate band

Husband and wife each have £3,000 chargeable at 20% and up to £20,000 at 25%.

Mortgage interest relief

The ceiling for relief is unchanged for 1994/95 at £30,000. The value of relief, however, has been reduced in a similar way to married couples allowance in that there is a maximum rate of relief of 20%. If the loan is in the name of one spouse only, that one gets all the relief due. If it is in the joint names of husband and wife it is allowed equally between them. However if both agree they can *jointly* ask for relief to be divided in any proportion that they wish.

Business losses

These are allowed only against the income of the person who incurs the loss. For example, a loss in the husband's business cannot be set against the wife's income from employment.

Joint income

In the case of joint ownership by husband and wife of assets which yield income - bank and building society accounts, shares, rented property, etc. - the Revenue will treat the income as arising equally to both and each will pay tax on one half of the income. If however the asset is owned in unequal shares or by one spouse only and the taxpayer can prove this, then the shares to be taxed can be adjusted accordingly if a joint declaration is made to the tax office setting out the facts.

General

Special rules apply in the year of marriage or separation and divorce and on the death of the husband or wife. Contact your local tax office for information.

Capital Gains Tax (individuals 1994/95)

Where an asset is disposed of, the first £5,800 of the gain is exempt from tax. In the case of husbands and wives, each has a £5,800 exemption so if the ownership of the assets is divided between them, it is possible to claim exemption on gains up to £11,600 jointly in the tax year. Any remaining gain is chargeable as though it were the top slice of the individual's income, therefore, according to his or her circumstances it might be charged at 20%, 25% or 40%. Here is an example:

TAXATION

Husband: He is in business and is liable at 40% on his profits. He has £7,000 of chargeable capital gains less the £5,800 exemption sum which leaves £1,200. This is taxed at 40% to produce a sum due to the Inland Revenue of £480.

Wife: She has no income but also has £7,000 of chargeable gains. Her exemption of £5,800 leaves a taxable sum of £1,200 and as she has no taxable income and the gain is less than the lower band it is charged at 20% making £240 to be paid in tax.

Retirement relief may be due on the disposal of certain business assets after the age of 55 (or before that date where retirement is due to proven ill health). The maximum relief against capital gains is £250,000 plus one half of the gains between £250,000 and £1,000,000. A businessman contemplating retirement, or sale of business when aged 55 or over should consult his accountant *before* taking any steps and *before* changing his working pattern (e.g. going part-time).

Business entertainment

No relief is due for expenditure on business entertainment or on gifts to customers, whether they are from this country or overseas. However, the cost of small trade gifts not exceeding £10 per person in value is still admissible provided that the gift advertises the business and does not consist of food or tobacco.

Construction industry

An uncertificated sub-contractor in the construction industry will suffer tax at the basic rate of 25%. Any overpayment on account of the 20% rate may be reclaimed at the end of the tax year.

Dates tax due

Income Tax 1994/95

Earned income (such as trading profits) 50% on 1 January 1995 and a further 50% on 1 July 1995. Unearned income (such as rents and interest) due 1 January 1995.

Capital Gains Tax

Tax on gains in year ended 5 April 1994 is due on 1 December 1994 (or within 30 days of the issue of the notice of assessment). Tax on gains arising in the current tax year are due on 1 December 1995.

Self-employed NIC rates (from 6 April 1994)

Class 2 Rate - £5.65 per week. If earnings are below £3,200 p.a. averaged over the year ask the DSS about 'small income exception'; details are in leaflet NI 27A.

Class 4 Rate - Business profits up to £6,490 p.a.: NIL. Profits between £6,490 and £22,360 p.a.: 7.3% of the profit. There is no charge on profits over £22,360 p.a. so the maximum amount of Class 4 contributions is £1,158.51. One half of the Class 4 contribution is deducted from the profits for income tax purposes. Class 4 contributions are collected by the Inland Revenue along with the income tax due.

Corporation Tax (years ended 31/3/94 and 31/3/95)

For the year ended 31/3/95 Corporation Tax is charged at 25% for profits up to £300,000. The ceiling for the previous year ended 31/3/94 was £250,000. Where the accounting date of a company is not 31/3/94, profits have to be apportioned on the time basis to the respective tax years. Profits exceeding £300,000 will be effectively charged at 35% up to £1,500,000 when the rate reduces to 33%. Companies can carry back trading losses for up to 3 years.

Capital allowances (depreciation) rates

Plant and machinery	25%	
Business motor cars - cost up to £12,000	25%	
- cost over £12,000	£3,000	(maximum)
Industrial buildings	4%	
Commercial and industrial buildings in Enterprise Zones	100%	

VAT

The standard rate is 17.5%. Lower limit for registration (from 11 March 1992) turnover per annum - £45,000.

Bad debts and VAT

Relief is available for debts over 6 months.

Chapter 4
Estimating

Pity the poor estimator! If a job goes well there will be a queue of agents, foremen, tradesmen, surveyors and buyers to take the credit. If a job loses money, however, the one person who will be left isolated to take the blame is the unsung hero in the construction industry - the estimator!

His art is highly dependent upon making a series of intelligent guesses to fill in the gaps of information not covered by the specification drawings and/or the bills of quantities. The quality of these guesses (or subjective judgements as the jargon would have it!) will often make the difference between a job making a profit or a loss.

It is possible, of course, to have too much information! There are many examples of a contractor carrying out the first phase of a contract but losing the second phase in open tender despite the fact of having the site set-up already there. The reason for the loss is usually due to having too much local knowledge gained from working on Phase 1 and including the cost of the known local risks in the second tender. Ignorance frequently prevails!

Apart from determining the contract sum at the tender stage, a properly prepared estimate can also be used for the following purposes:

1. Calculation of bonus targets;

2. Preparation of material schedules;

3. Analysis of anticipated and actual costs;

4 Production of a programme;

5. Preparation of monthly valuation and the final account.

This book, however, is only concerned with the preparation of estimates at the tender stage. It should be recognized that estimating is a very imprecise art. This often surprises people outside the construction industry who imagine it to be '...merely a matter of counting bricks and pricing them' as someone once said to me. If only it was!

Even taking this comment at face value shows how inaccurate this view is. There are many different kinds of bricks all with a different value. The cost of sand and cement for the mortar varies from supplier to supplier as does the hire cost of a mixer. The labour costs are the most likely to show the greatest variation.

No two men work at the same pace and produce the same volume of work. So even an uncomplicated task such as building a wall can produce significantly different estimates. When it comes to more complicated work, the likelihood of different estimates producing wide variations in their tender increases proportionately.

The main divisions in a properly prepared estimate are:

1. Own work which can be sub-divided into
   (a) Labour
   (b) Materials
   (c) Plant;

2. Work to be sub-let to sub-contractors;

3. Site overheads;

4. Office overheads;

5. Profit.

When an enquiry is received a contractor should decide very quickly which parts of the work he would sub-let if he was awarded the contract. He should then send the relevant extracts from the enquiry documents (specification/bills of quantities/drawings) to two or three sub-contractors whilst he is preparing the remainder of the estimate.

Labour

An experienced contractor should know the net charge-out rate of the men he directly employs. The basic labour rate on which the information in this book is based is set out at the beginning of Chapter 5. Every contractor will probably have his own 'customized' version but the main items such as NI contribution, overtime payments, bonus, etc., must be included.

The particular circumstances of the job must also be considered. The work may have to be carried out outside normal trading hours or done in unpleasant working conditions - both these situations would produce higher labour costs.

ESTIMATING

Materials

Although it is usually possible to get a better price from another materials supplier, a contractor should weigh up whether it is worth spending a lot of his time investigating other sources of supply if the savings are marginal.

A supplier who delivers on time (including the occasional small item when necessary) and replaces defective materials without query is probably worth supporting even if his prices are slightly higher than some of his competitors. It is still worth checking other prices now and then, however, to make sure that they are not too far out of line; but reliability and service have a real value and should be recognized.

The most difficult aspect of pricing materials is making a realistic assessment for waste and theft. Materials which are bought in bulk and only part of the order is allocated to any particular job or operation usually attract the highest waste percentage which can be as high as 15 to 20%. Specialized 'one-off' items such as cylinders, heaters and the like are usually better looked after and the waste factor could be as low as 1%.

Theft is another problem and a prudent contractor should make some allowance in his estimates to cover for this frustrating part of the industry.

Plant

Most small contractors hire in plant as necessary so the estimator need only assess the time the plant will be required because the hire rate will be easily established. One point to watch out for is the question of minimum hire periods.

Even if a piece of equipment is only required for one hour, the full cost of a day's hire must be included in the rates if that is the minimum hire period that can be obtained.

Site overheads

The range of overheads to be provided on site will vary widely depending on the size and nature of the job. There are two main types of overheads - fixed and time-related. Examples of these will clearly reveal the difference between them.

Fixed overheads are those of a non-recurring nature such as the establishment of the site huts. The cost of this operation - say £500 - must be allowed for whether the job lasts 4 weeks or 104 weeks and is a fixed sum.

Time-related overheads however are totally linked to the time the facility or service is required. In the case of site hutting the estimator should allow a weekly cost for the anticipated contract period plus whatever time he thinks the huts will be needed for carrying out maintenance work.

The following should act as a check list for both fixed and time-related overheads:

Site supervision
Site accommodation
Lighting
Water
Safety, health and welfare
Removal of rubbish
Cleaning
Drying out
Protection of work
Security
Small tools
Insurances
Travelling time and fares
Scaffolding
Temporary services
Temporary fencing and screens

Office overheads

All contractors should give careful thought to the cost of their office overheads and in particular should be aware of the relationship, expressed as a percentage, between these costs and turnover.

The following is an example of a small contractor's overheads. This list is not meant to be exhaustive.

			£
Rent		(52 x £50)	2,600
Printing, stationery, etc.			150
Telephone		(12 x £50)	600
Van	HP	(12 x £200)	2,400
	Insurance		400
	Tax		100
	Petrol	(52 x £30)	1,560
	Repairs	say	500
Carried forward		£	8,310

ESTIMATING

Brought forward		£	8,310
Part-time clerk/typist (52 x 30 hours x £3.50)			5,460
Photocopier - lease	(52 x £25)		1,300
Word-processor - lease	(52 x £20)		1,040
Advertising	(52 x £20)		1,040
Membership of professional body			150
Sundries			1,000
		£	18,300

Let us assume that the firm's turnover (that is, the total amount billed for the previous year) is £100,000. A simple calculation of £18,300 x 100 ÷ £100,000 shows that the relationship of overheads to turnover is 18.3%. I have set out a table below showing how this percentage varies when the level of the overheads or the turnover changes.

Overheads £	Turnover £	%
18,300	100,000	18.30
18,300	80,000	22.90
18,300	120,000	15.25
24,000	100,000	24.00
30,000	80,000	37.50
12,000	120,000	10.00

It is important to know the current percentage so it can be added to each quotation (although not necessarily shown). Ask your accountant for the cost of the overheads for last year and assess the percentage to be added to the costs. Unless something dramatic happens mid-year this figure needs only be examined annually. An example which would affect the overheads total could be the employment of a non-working supervisor or the purchase of an extra vehicle. A cancellation of a project or an unexpected increase in work would also affect the turnover figure.

It can be seen, therefore, that although working on a notional percentage is acceptable you should keep your eye open for any circumstances which may make a significant alteration to it.

Profit

Profit is the difference between income and cost. Some newly established contractors find it puzzling that although they are certain that their work is profitable, it is not reflected in the bank balance! The simple explanation for this apparent paradox is that as soon as it is earned the profit is re-used to finance the next stage of work or the next job. It only appears as a tangible asset when either trading stops or after a series of profitable jobs are completed and paid for.

Expansion of the business (and it is very difficult for a successful contractor not to expand) will further delay the surfacing of the profit and even a well planned expansion programme will usually result in an increased overdraft. The profit is still in the business of course, although not in the form of cash in the bank but in debtors' accounts and work in progress.

Chapter 5
Rates for Measured Work

Generally

The rates contained in this chapter apply to contracts ranging from minor works to those up to £50,000 in value. The rates are based upon national average prices and the following adjustments should be made for working in different regions

Scotland		+ 10%
Wales		- 4%
Northern Ireland		- 18%
England		
	South West	- 8%
	South East	+ 5%
	Home Counties	+ 7%
	Inner London	+ 20%
	Outer London	+ 12%
	East Anglia	+ 2%
	Midlands	- 3%
	North West	+ 2%
	North East	- 3%

The rates are calculated using material and plant costs current in the second quarter of 1994 and the wage rates set by the following organizations:

General building   Building and Allied Trades Joint Industry Council (BATJIC) from October 1993

Plumbing   Joint Industry Board for Plumbing, Mechanical and Engineering Services from 4 April 1994

RATES FOR MEASURED WORK

The rates are exclusive of VAT which may be chargeable depending upon the nature of the work and are divided into their component parts so that adjustments can easily be made if necessary.

Unit

This column gives the unit in which the item is normally measured.

Item descriptions

Each item includes all the work normally associated with that particular item even if it is not expressly stated. Where the words *Add to the previous prices for* appear in the text, the costs which appear opposite the item represent the *additional* cost of replacing the component in the main description with the item described. Where the basic price of an item is expressed by the letters BP it represents the purchase cost net of any discounts, which the builder could expect to pay over the trade counter.

Labour hours

The time taken for the fixing of each of the items is expressed as a decimal fraction of an hour; therefore 0.50 Labour hours equals thirty minutes. The times are average and allow for unloading, distributing and fixing in position.

Net labour

This column gives the total labour cost of supplying and fixing the item concerned. It is calculated by multiplying the total labour hours by the hourly rate.

General Building Work

		Craftsman £		Labourer £
Annual wages 1911 hours	at £ 4.44	8,484.84	at £ 3.82	7,300.02
Non-productive overtime 89 hours	at £ 4.44	395.16	at £ 3.82	339.98
Sick pay 5 days	at £13.37	66.85	at £13.37	66.85
Carried forward		£ 8,946.85		7,706.85

42

RATES FOR MEASURED WORK

Brought forward		£ 8,946.85		7,706.85
Public holidays 8 days	at £35.52	284.16	at £30.56	244.48
Tool allowance 49 weeks	at £ 1.03	50.47		-
Bonus say 30%	at £4.44	2,545.05	at £3.82	2,190.01
		£11,826.53		£10,141.34
Employers NIC		1,183.64		1,015.27
CITB Levy at 0.25%		29.56		25.35
Holidays with pay 47 weeks at £16.14		758.58		758.58
Retirement benefit 52 weeks	at £1.27	66.04		66.04
Death benefit 12 months at £3.61		43.32		43.32
		£13,907.67		£12,049.90
Severance pay say 1.5%		208.62		180.75
Employer's Liability and Third Party Insurance say 2%		278.15		241.00
TOTAL ANNUAL COST		£14,394.44		£12,471.65
COST PER HOUR divided by 1822 (1911 less 89 = 1822 productive hours)		£7.81		£6.76
		say £8.00		£7.00

Plumber

The Net labour column gives the total labour cost of supplying and fixing the item concerned. It is calculated by multiplying the total Plumber hours by the hourly rate. The all-in rate is calculated in accordance with the rates of wages agreed and published by the Joint Industry Board for Plumbing, Mechanical

RATES FOR MEASURED WORK

Engineering Services in England and Wales as follows:

	Trained Plumber	Advanced Plumber	Technical Plumber
Basic rate (from 4 April 1994)	£4.96	£5.49	£6.20
Basic rate x total hours worked (1910 hrs)	9,473.60	10,485.90	11,842.00
Non-productive overtime (50 hrs)	248.00	274.50	310.00
Public holidays (60 hrs)	297.60	329.40	372.00
Sub-total (1)	10,019.20	11,089.80	12,524.00
National Insurance (based on sub-total 1)	764.40	861.50	956.00
Annual holiday with pay and sickness benefit 52 weeks at: £17.23 (Trained Plumber) £18.76 (Advanced Plumber) £21.01 (Technical Plumber)	859.96	975.52	1,092.52
Industry pension scheme 6.5% of sub-total (1)	651.25	720.84	814.06
Tool money 45.2 weeks at £2.24	101.25	101.25	101.25
CITB Levy (0.25%)	25.05	27.72	31.31
Sub-total	12,421.11	13,776.63	15,519.14
Severance pay 1.5%	186.32	206.65	232.79
Employer's liability and third party insurance 2%	248.42	275.53	310.38
TOTAL ANNUAL COST	12,855.85	14,258.81	16,062.31
COST PER HOUR divided by 1822 (1911 less 89 = 1822 productive hours)	£6.89	£7.63	£8.59

RATES FOR MEASURED WORK

The hourly rate is based therefore upon 2 advanced plumbers @ £7.63 = £15.26 plus 1 third year apprentice @ £3.96 = £19.22 ÷ two principal operatives = £9.61 say £9.60.

The following hourly rates are also used in the book:

Brick and block laying gang:

2 craftsmen	@ £8.00	=	16.00
1 labourer	@ £7.00	=	7.00
			£23.01
÷ 2 craftsmen		=	£11.50

Plastering and finishing gang

3 craftsmen	@ £8.00	=	24.00
1 labourer	@ £7.00	=	7.00
			31.00
÷ 3 craftsmen		=	£10.33

All other craftsmen working alone = £ 8.00

Net material

The material costs are set out in the front of the work sections. Generally speaking, the costs reflect the manufacturers' list prices and exclude trade discounts which are usually negotiated by each individual contractor depending upon his creditworthiness, turnover and length of association with the supplier. The prices are based upon the purchase of manufacturers' standard packs.

Overheads/profit

Overhead charges are those costs which are not directly linked to any particular contract. They are usually expressed as a percentage of the net cost of labour and materials and are calculated by dividing the total projected annual overhead cost by the projected net turnover and are explained in more detail in Chapter 4.

The percentage addition for overheads and profit is taken at 15%.

Total

The total column is the sum of the labour, material and overheads/profit columns.

Plant

The rates and prices contained within this book include for working off ladders and scaffoldings. The hire rates for small tools and plant are given in Chapter 6 and an allowance should be made in the preliminaries for the cost.

C

DEMOLITION/ALTERATION/RENOVATION

Spon's Building Costs Guide for Educational Premises

Edited by **Tweeds**, Chartered Quantity Surveyors, UK

"this unique book [is] an invaluable information source." - *The Groundsman*

This book is for those head teachers and governors who have recently taken on the responsibility of handling building work under the Local Management of Schools Scheme. The book sets out common sense rules on the appointment of contractors, monitoring and settling their final accounts. In particular it provides information on the current costs of building work together with details on budget estimating procedures. Ancillary information on professional fees, VAT, a glossary of building terms and a section of model letters on a small painting contract are also included.

Contents include: Preface. Introduction. **Part One.** Contract administration: records and accounting. Choosing a contractor. Obtaining quotations. Comparing tenders. Variations. Payments. VAT. '714' scheme. Safety. Contracts: generally. Agreement for minor building works. Insurances. **Part Two.** Measured rates for new work. Internal: floors. Walls. Ceilings. Carpentry and joinery. Plumbing. Wallpapering. External: roofing. Rainwater goods. Precast concrete kerbs and channels. Hardcore sub-bases. Insitu concrete bases. Drains and manholes. Measured rates for repairs. Internal: spot items. Ceilings. Joinery. Macadam pavings. Fencing and gates. Measured rates for cleaning and maintenance: cleaning. Measured rates for equipment: classroom furniture. Blackboards. Sports and playground equipment. Flagpoles. Budget estimating: square metre prices. Elemental costs. Composite rates. **Part Three.** Professional fees. Standard letters. Glossary. Useful addresses. Index.

January 1994: 234x156: 256pp
Hardback: 0-419-18860-6: £35.00

For further information and to order please contact: The Promotion Dept., E & F N Spon, 2-6 Boundary Row, London SE1 8HN
Tel: 071 865 0066 Fax: 071 522 9623

E & F N Spon
An imprint of Chapman & Hall

	Unit	Material supply (£)	Material waste (%)	Total (£)

C DEMOLITION/ALTERATION/ RENOVATION

The cost and waste allowances for materials in this section can be assessed from other relevant parts of the book where they appear in detail

RATES FOR MEASURED WORK

	Unit	Labour hours	Net labour (£)	Net material (£)	O'heads /profit (£)	Total (£)

C DEMOLITION/ALTERATION/RENOVATE

Labour rate 7.00 per hour

There are a number of factors which directly affect the cost of the demolition of buildings, e.g. access, the residual value of the arisings, proximity of other buildings, etc.

Because of this the following rates should be treated with caution and regarded as indicative only.

GENERAL DEMOLITION

Demolish existing to ground level

Brick outbuildings

	Unit	Labour hours	Net labour (£)	Net material (£)	O'heads /profit (£)	Total (£)
single storey detached	m3	0.00	0.00	8.66	2.58	19.76
single storey attached	m3	0.00	0.00	9.66	1.45	11.11
two storey attached	m3	0.00	0.00	10.66	1.60	12.26

Break up and remove, load into skips or lorries

Unreinforced concrete slabs

	Unit	Labour hours	Net labour (£)	Net material (£)	O'heads /profit (£)	Total (£)
100mm thick	m2	0.75	5.25	0.00	0.79	6.04
150mm thick	m2	1.00	7.00	0.00	1.05	8.05
250mm thick	m2	2.00	14.00	0.00	2.10	16.10

Reinforced concrete slabs

	Unit	Labour hours	Net labour (£)	Net material (£)	O'heads /profit (£)	Total (£)
100mm thick	m2	2.00	14.00	0.00	2.10	16.10
150mm thick	m2	2.50	17.50	0.00	2.62	20.12
250mm thick	m2	3.00	21.00	0.00	3.15	24.15

50

DEMOLISHING STRUCTURES

	Unit	Labour hours	Net labour (£)	Net material (£)	O'heads /profit (£)	Total (£)
Reinforced suspended concrete slabs						
150mm thick	m2	2.75	19.25	0.00	2.89	22.14
200mm thick	m2	3.00	21.00	0.00	3.15	24.15
250mm thick	m2	3.50	24.50	0.00	3.67	28.17
Reinforced concrete beams						
200 x 300mm	m2	1.20	8.40	0.00	1.26	9.66
250 x 350mm	m2	1.75	12.25	0.00	1.84	14.09
300 x 400mm	m2	2.40	16.80	0.00	2.52	19.32
350 x 500mm	m2	3.50	24.50	0.00	3.67	28.17
Reinforced concrete columns						
250 x 250mm	m	1.25	8.75	0.00	1.31	10.06
300 x 300mm	m	1.80	12.60	0.00	1.89	14.49
300 x 450mm	m	2.70	18.90	0.00	2.83	21.73
400 x 600mm	m	4.80	33.60	0.00	5.04	38.64
Reinforced concrete staircase 2.70m rise and 2.30m going (balustrades taken down elsewhere) straight flight						
1000mm wide	nr	8.00	56.00	0.00	8.40	64.40
1250mm wide	nr	10.00	70.00	0.00	10.50	80.50
Brick walls in cement mortar						
half brick thick	m2	0.50	3.50	0.00	0.53	4.03
one brick thick	m2	0.80	5.60	0.00	0.84	6.44
one and a half brick thick	m2	1.50	10.50	0.00	1.57	12.07
Brick walls in cement lime mortar						
half brick thick	m2	0.40	2.80	0.00	0.42	3.22
one brick thick	m2	0.75	5.25	0.00	0.79	6.04
one and a half brick thick	m2	1.00	7.00	0.00	1.05	8.05

RATES FOR MEASURED WORK

General demolition (cont'd)	Unit	Labour hours	Net labour (£)	Net material (£)	O'heads /profit (£)	Total (£)
Taking down with care, in cement lime mortar						
half brick thick	m2	0.60	4.20	0.00	0.63	4.83
one brick thick	m2	1.10	7.70	0.00	1.16	8.86
one and a half brick thick	m2	1.50	10.50	0.00	1.57	12.07
Taking down with care, set aside bricks for reuse after cleaning						
half brick thick	m2	1.00	7.00	0.00	1.05	8.05
one brick thick	m2	1.50	10.50	0.00	1.57	12.07
one and a half brick thick	m2	2.20	15.40	0.00	2.31	17.71
Take out fireplace surround and hearth, remove, load into skip or lorries						
1.50m wide x 1.20m high in tiled concrete	nr	1.00	7.00	0.00	1.05	8.05
Hack face of existing brickwork and rake out joints for new plaster	m2	0.50	3.50	0.00	0.53	4.03
Hack off plaster on existing brickwork and prepare to receive new plaster	m2	0.75	5.25	0.00	0.79	6.04
Hack off cement and sand render on existing brickwork and prepare to receive new render	m2	0.85	5.95	0.00	0.89	6.84

	Unit	Labour hours	Net labour (£)	Net material (£)	O'heads /profit (£)	Total (£)

C20 ALTERATIONS - SPOT ITEMS

Labour rate 7.00 per hour

FORMING OPENINGS

Cut opening for lintel or beam
65mm deep through existing wall

half brick wall	m	0.35	2.45	0.00	0.37	2.82
one brick wall	m	0.55	3.85	0.00	0.58	4.43
one and a half brick wall	m	0.69	4.83	0.00	0.72	5.55
100mm blockwork	m	0.28	1.96	0.00	0.29	2.25
150mm blockwork	m	0.31	2.17	0.00	0.33	2.50
200mm blockwork	m	0.37	2.59	0.00	0.39	2.98
250mm blockwork	m	0.41	2.87	0.00	0.43	3.30

Cut opening for lintel or beam
90mm deep through existing wall

half brick wall	m	0.40	2.80	0.00	0.42	3.22
one brick wall	m	0.62	4.34	0.00	0.65	4.99
one and a half brick wall	m	0.75	5.25	0.00	0.79	6.04
100mm blockwork	m	0.32	2.24	0.00	0.34	2.58
150mm blockwork	m	0.35	2.45	0.00	0.37	2.82
200mm blockwork	m	0.42	2.94	0.00	0.44	3.38
250mm blockwork	m	0.45	3.15	0.00	0.47	3.62

Cut opening for lintel or beam
150mm deep through existing wall

half brick wall	m	0.55	3.85	0.00	0.58	4.43
one brick wall	m	0.88	6.16	0.00	0.92	7.08
one and a half brick wall	m	1.07	7.49	0.00	1.12	8.61
100mm blockwork	m	0.42	2.94	0.00	0.44	3.38
150mm blockwork	m	0.52	3.64	0.00	0.55	4.19
200mm blockwork	m	0.60	4.20	0.00	0.63	4.83
250mm blockwork	m	0.64	4.48	0.00	0.67	5.15

Forming openings (cont'd)	Unit	Labour hours	Net labour (£)	Net material (£)	O'heads /profit (£)	Total (£)
Cut opening for door or window through existing wall						
half brick wall	m2	2.10	14.70	0.00	2.20	16.90
one brick wall	m2	2.95	20.65	0.00	3.10	23.75
one and a half brick wall	m2	3.82	26.74	0.00	4.01	30.75
100mm blockwork	m2	1.54	10.78	0.00	1.62	12.40
150mm blockwork	m2	1.77	12.39	0.00	1.86	14.25
200mm blockwork	m2	2.35	16.45	0.00	2.47	18.92
250mm blockwork	m2	2.55	17.85	0.00	2.68	20.53

Form opening for door and frame through internal wall plastered both sides, take off skirting, cut out for new lintel, quoin up jambs, make good plaster both sides and into reveals, make good ends of skirting both sides and into reveal up to new frame

900 x 2000mm	nr	18.40	128.80	38.01	25.02	191.83

The following rates were used to compile the above and similar work

cut opening through half brick wall	m2	2.00	14.00	0.00	2.10	16.10
cut opening through one brick wall	m2	3.00	21.00	0.00	3.15	24.15
Catnic lintel to half brick wall	nr	1.00	7.00	18.06	3.76	28.82
Catnic lintel to one brick wall	nr	1.50	10.50	28.66	5.87	45.03
precast concrete lintel to one brick wall	nr	1.75	12.25	9.44	3.25	24.94
wedge and pin up over lintel in half brick wall	m	0.50	3.50	0.00	0.53	4.03
wedge and pin up over lintel in one brick wall	m	0.50	3.50	0.00	0.53	4.03
quoin up half brick jambs	m	0.30	2.10	0.00	0.32	2.42
quoin up one brick jambs	m	0.60	4.20	0.00	0.63	4.83
make good plaster to wall and reveal up to new frame	m	0.75	5.25	0.37	0.84	6.46
make good plaster to wall	m	0.35	2.45	0.37	0.42	3.24

ALTERATIONS - SPOT ITEMS

	Unit	Labour hours	Net labour (£)	Net material (£)	O'heads /profit (£)	Total (£)
make good skirting up to new frame including mitres using old skirting removed	nr	0. 40	2. 80	0. 00	0. 42	3. 22
extend existing floor boarding through opening, provide bearers	nr	2. 50	17. 50	4. 31	3. 27	25. 08
lay cement and sand screed through opening for floor finish to match existing	m2	1. 50	10. 50	1. 26	1. 76	13. 52

Form opening for window through
external wall plastered one side
faced other side, insert lintel,
quoin up jambs, make good facings
to margin, make good plaster,
lead cored damp-proof course over
lintel and at closed jambs, lay
slates at cill in cement mortar

	Unit	Labour hours	Net labour (£)	Net material (£)	O'heads /profit (£)	Total (£)
1200 x 850mm	nr	30. 00	210. 00	60. 79	40. 62	311. 41

The following rates were used to
compile the above and similar work

	Unit	Labour hours	Net labour (£)	Net material (£)	O'heads /profit (£)	Total (£)
cut opening through cavity wall 275mm thick with 110mm blockwork inner skin	m2	3. 50	24. 50	0. 00	3. 67	28. 17
cut opening through cavity wall 275mm thick with half brick inner skin	m2	4. 00	28. 00	0. 00	4. 20	32. 20
Catnic lintel to 275mm cavity wall up to 2100mm long	nr	2. 00	14. 00	49. 96	9. 59	73. 55
Catnic lintel over 2100mm long	nr	2. 50	17. 50	58. 85	11. 45	87. 80
Precast concrete lintel 100 x 150mm to skin of cavity wall	m	2. 50	17. 50	9. 44	4. 04	30. 98
wedge and pin up over lintel in half brick or block wall	m	1. 00	7. 00	0. 00	1. 05	8. 05
quoin up half brick wall in facings	m	0. 30	2. 10	2. 02	0. 62	4. 74
quoin up block wall internally	m	0. 10	0. 70	1. 33	0. 30	2. 33

RATES FOR MEASURED WORK

Forming openings (cont'd)	Unit	Labour hours	Net labour (£)	Net material (£)	O'heads /profit (£)	Total (£)
brick on end arch in facing bricks to match existing faced work including taking out three courses of existing wall and inserting mild steel arch bar	m	1.20	8.40	12.31	3.11	23.82
Form opening for door and frame through external wall, door head at height of old window head, cut out brickwork under cill, quoin up jambs and face margins, retain existing lintel and arch over, close cavity each side, construct new concrete step at threshold	nr	30.00	210.00	60.79	40.62	311.41
The following rates were used to compile the above and similar work						
cut opening below window through 275mm cavity wall	m2	4.00	28.00	0.00	4.20	32.20
cut opening below window through one and half brick wall	m2	5.00	35.00	0.00	5.25	40.25
quoin up half brick jambs in red facing bricks to match existing	m	1.50	10.50	2.02	1.88	14.40
make good inner skin, close cavity	m	0.50	3.50	0.00	0.53	4.03
form fitted ends of existing skirting up to new frame	nr	0.50	3.50	1.17	0.70	5.37

ALTERATIONS - SPOT ITEMS

	Unit	Labour hours	Net labour (£)	Net material (£)	O'heads /profit (£)	Total (£)
FILLING OPENINGS						

Fill opening where door removed
from internal wall of 100mm
concrete blocks, bond blockwork
to existing at jambs, wedge and
pin at head, plaster both sides
to match existing and extend
skirting both sides to match
existing

	Unit	Labour hours	Net labour (£)	Net material (£)	O'heads /profit (£)	Total (£)
900 x 2000mm high	nr	15.50	108.50	21.20	19.45	149.15

Rates used for compiling the above
and similar work

	Unit	Labour hours	Net labour (£)	Net material (£)	O'heads /profit (£)	Total (£)
75mm blockwork in filling	m2	1.00	7.00	6.20	1.98	15.18
100mm blockwork in filling	m2	1.25	8.75	6.63	2.31	17.69
112mm brickwork in filling	m2	2.00	14.00	10.07	3.61	27.68
bonding blockwork to existing 75mm thick	m	0.50	3.50	0.00	0.53	4.03
bonding blockwork to existing 100mm thick	m	0.60	4.20	0.00	0.63	4.83
two coat plaster to filling openings	m2	1.30	9.10	1.05	1.52	11.67
softwood skirting to match existing	m	0.50	3.50	0.00	0.53	4.03

Fill opening where door removed
from one brick wall faced
externally, remove existing lintel
and arch, provide new damp-proof
course, bond new brickwork to
existing, wedge and pin up at
head, make good plaster and extend
skirting to match existing

	Unit	Labour hours	Net labour (£)	Net material (£)	O'heads /profit (£)	Total (£)
900 x 2000mm high	nr	20.25	141.75	62.79	30.68	235.22

RATES FOR MEASURED WORK

Filling openings (cont'd)	Unit	Labour hours	Net labour (£)	Net material (£)	O'heads /profit (£)	Total (£)
one brick filling in common bricks faced one side to match existing	m2	5.00	35.00	28.55	9.53	73.08
one brick filling in common bricks finished with fair face internally and facings externally	m2	4.00	28.00	28.86	8.53	65.39
bituminous felt damp-proof course with lead core 225mm wide	m	0.75	5.25	3.16	0.21	8.62
wedge and pin up brickwork at head of opening	m	1.00	7.00	0.00	1.05	8.05
cutting toothings for bonding one brick filling to existing brickwork at jambs	m	1.50	10.50	0.00	1.57	12.07
cut away existing concrete step	nr	2.00	14.00	0.00	2.10	16.10

ALTERATIONS

Rates include wheeling and depositing in skip

Pull down concrete walls

	Unit	Labour hours	Net labour	Net material	O'heads /profit	Total
100mm thick	m2	1.35	9.45	6.50	2.39	18.34
150mm thick	m2	1.90	13.30	9.29	3.39	25.98
200mm thick	m2	2.85	19.95	10.94	4.63	35.52

Pull down reinforced concrete walls

	Unit	Labour hours	Net labour	Net material	O'heads /profit	Total
100mm thick	m2	1.60	11.20	8.90	3.01	23.11
150mm thick	m2	2.24	15.68	12.93	4.29	32.90
200mm thick	m2	3.15	22.05	16.55	5.79	44.39

Break up concrete ground floor slabs

	Unit	Labour hours	Net labour	Net material	O'heads /profit	Total
100mm thick	m2	0.64	4.48	5.16	1.45	11.09
150mm thick	m2	0.95	6.65	7.00	2.05	15.70

ALTERATIONS - SPOT ITEMS

	Unit	Labour hours	Net labour (£)	Net material (£)	O'heads /profit (£)	Total (£)
Break up reinforced concrete ground floor slab						
100mm thick	m2	0.98	6.86	7.07	2.09	16.02
150mm thick	m2	1.15	8.05	9.69	2.66	20.40
Break up suspended concrete floor slabs						
150mm thick	m2	2.26	15.82	5.25	3.16	24.23
200mm thick.	m2	3.42	23.94	7.08	4.65	35.67
Break up suspended reinforced concrete floor slab						
150mm thick	m2	3.20	22.40	7.14	4.43	33.97
200mm thick	m2	3.90	27.30	9.35	5.50	42.15
Form opening through reinforced concrete wall or suspended slab						
150mm thick	m2	2.80	19.60	6.53	3.92	30.05
200mm thick	m2	3.65	25.55	8.72	5.14	39.41
Cut hole in 100mm thick reinforced concrete and make good for						
small pipe	nr	0.75	5.25	5.36	1.59	12.20
large pipe	nr	1.12	7.84	2.48	1.55	11.87
Cut hole in 150mm thick reinforced concrete and make good for						
small pipe	nr	0.85	5.95	2.73	1.30	9.98
large pipe	nr	1.30	9.10	3.52	1.89	14.51
Cut out and enlarge crack in concrete, fill with mortar	m	0.65	4.55	1.49	0.91	6.95

	Unit	Labour hours	Net labour (£)	Net material (£)	O'heads /profit (£)	Total (£)

REPAIRS

The items listed here are for work where there is no requirement for supports during the pulling down. Breaking out openings where supports are required is dealt with separately. All rates include for placing in skips.

If the materials being demolished are required for reuse 40% should be added to total rate to allow for the extra cost in carefully taking down and laying aside for reuse.

Brickwork

Labour rate 7.00 per hour

Pull down brick walls in gauged mortar, thickness

half brick	m2	0.80	5.60	0.00	0.84	6.44
one brick	m2	1.35	9.45	0.00	1.42	10.87
one and a half brick	m2	1.95	13.65	0.00	2.05	15.70
two brick	m2	2.48	17.36	0.00	2.60	19.96

Pull down brick walls in cement mortar, thickness

half brick	m2	1.32	9.24	0.00	1.39	10.63
one brick	m2	1.80	12.60	0.00	1.89	14.49
one and a half brick	m2	2.50	17.50	0.00	2.62	20.12
two brick	m2	3.31	23.17	0.00	3.48	26.65

Take out timber or tiled fireplace surround, size

1000 x 900mm	nr	0.60	4.20	0.00	0.63	4.83
1200 x 900mm	nr	0.70	4.90	0.00	0.73	5.63

ALTERATIONS - SPOT ITEMS

	Unit	Labour hours	Net labour (£)	Net material (£)	O'heads /profit (£)	Total (£)
Take out tiled concrete hearth, size						
1000 x 450mm	nr	0.75	5.25	0.00	0.79	6.04
1200 x 600mm	nr	0.92	6.44	0.00	0.97	7.41
Labour rate 8.00 per hour						
Take down brick chimney to below roof level, cap with slates in cement mortar (1:3), size						
450 x 450 x 900mm high	nr	9.40	75.20	17.42	2.32	94.94
450 x 900 x 1200mm high	nr	13.60	108.80	26.36	3.38	138.54
Cut back brick chimney breast projections, size						
half brick thick	m2	2.18	17.44	4.49	3.29	25.22
one brick thick	m2	2.78	22.24	6.63	4.33	33.20
one and a half brick thick	m2	4.10	32.80	10.96	6.56	50.32
Make good newly exposed end of wall in common brickwork to form new jambs to receive plaster, thickness						
half brick wall	m	0.42	3.36	2.92	0.94	7.22
one brick wall	m	0.69	5.52	4.74	1.54	11.80
one and a half brick wall	m	0.95	7.60	6.31	2.09	16.00
Make good newly exposed end of wall in common brickwork to form new jambs, fair faced and flush pointed						
half brick wall	m	0.50	4.00	3.54	1.13	8.67
one brick wall	m	0.82	6.56	6.05	1.89	14.50
one and a half brick wall	m	1.21	9.68	7.89	2.64	20.21

61

RATES FOR MEASURED WORK

Repairs (cont'd)	Unit	Labour hours	Net labour (£)	Net material (£)	O'heads /profit (£)	Total (£)
Blockwork						
Labour rate 8.00 per hour						
Pull down block walls in gauged mortar (1:1:6), thickness						
100mm	m2	0.87	6.96	0.00	1.04	8.00
150mm	m2	1.10	8.80	0.00	1.32	10.12
200mm	m2	1.32	10.56	0.00	1.58	12.14
250mm	m2	1.63	13.04	0.00	1.96	15.00
Pull down block walls in cement mortar (1:3), thickness						
100mm	m2	1.08	8.64	0.00	1.30	9.94
150mm	m2	1.32	10.56	0.00	1.58	12.14
200mm	m2	1.60	12.80	0.00	1.92	14.72
250mm	m2	2.42	19.36	0.00	2.90	22.26
Rubble walling						
Labour rate 8.00 per hour						
Pull down rubble walling in gauged mortar, thickness						
200mm	m2	0.55	4.40	0.00	0.66	5.06
300mm	m2	0.65	5.20	0.00	0.78	5.98
450mm	m2	0.80	6.40	0.00	0.96	7.36
Carpentry						
Labour rate 8.00 per hour						
Take down, cut out or demolish and load into skips						
Structural timbers						
50 x 100mm	m	0.10	0.80	0.00	0.12	0.92
50 x 150mm	m	0.11	0.88	0.00	0.13	1.01
75 x 100mm	m	0.14	1.12	0.00	0.17	1.29

ALTERATIONS - SPOT ITEMS

	Unit	Labour hours	Net labour (£)	Net material (£)	O'heads /profit (£)	Total (£)
75 x 150mm	m	0.17	1.36	0.00	0.20	1.56
100 x 150mm	m	0.20	1.60	0.00	0.24	1.84
100 x 200mm	m	0.22	1.76	0.00	0.26	2.02
Roof boarding	m2	0.28	2.24	0.00	0.34	2.58
Floor boarding	m2	0.22	1.76	0.00	0.26	2.02
Stud partition plastered both sides	m2	0.85	6.80	0.00	1.02	7.82
Skirting and grounds						
100mm high	m	0.07	0.56	0.00	0.08	0.64
150mm high	m	0.08	0.64	0.00	0.10	0.74
250mm high	m	0.10	0.80	0.00	0.12	0.92
Rails						
50mm high	m	0.05	0.40	0.00	0.06	0.46
75mm high	m	0.06	0.48	0.00	0.07	0.55
100mm high	m	0.07	0.56	0.00	0.08	0.64
Standard kitchen fittings						
wall cupboards	nr	0.25	2.00	0.00	0.30	2.30
floor units	nr	0.20	1.60	0.00	0.24	1.84
sink units	nr	0.25	2.00	0.00	0.30	2.30
staircase 900mm wide, single						
straight flight	nr	3.70	29.60	0.00	4.44	34.04
landing	nr	1.20	9.60	0.00	1.44	11.04
internal door and lining	nr	0.35	2.80	0.00	0.42	3.22
external door and frame	nr	0.45	3.60	0.00	0.54	4.14
Casement window						
1200 x 900mm	nr	0.50	4.00	0.00	0.60	4.60
1800 x 900mm	nr	0.55	4.40	0.00	0.66	5.06
Sash window						
900 x 1500mm	nr	0.75	6.00	0.00	0.90	6.90
1200 x 1800mm	nr	0.90	7.20	0.00	1.08	8.28
porch canopy	nr	1.20	9.60	0.00	1.44	11.04
Ironmongery						
bolt	nr	0.20	1.60	0.00	0.24	1.84
deadlock	nr	0.30	2.40	0.00	0.36	2.76
mortice lock	nr	0.40	3.20	0.00	0.48	3.68

RATES FOR MEASURED WORK

Repairs (cont'd)	Unit	Labour hours	Net labour (£)	Net material (£)	O'heads /profit (£)	Total (£)
mortice latch	nr	0.40	3.20	0.00	0.48	3.68
cylinder lock	nr	0.25	2.00	0.00	0.30	2.30
door closer	nr	0.30	2.40	0.00	0.36	2.76
casement stay	nr	0.15	1.20	0.00	0.18	1.38
casement fastener	nr	0.15	1.20	0.00	0.18	1.38
toilet roll holder	nr	0.15	1.20	0.00	0.18	1.38
shelf bracket	nr	0.10	0.80	0.00	0.12	0.92
Cut out defective joists or rafters and replace with new						
50 x 75mm	m	0.35	2.80	0.85	0.55	4.20
50 x 100mm	m	0.48	3.84	1.05	0.73	5.62
50 x 150mm	m	0.55	4.40	1.74	0.92	7.06
75 x 100mm	m	0.60	4.80	2.08	1.03	7.91
75 x 150mm	m	0.82	6.56	2.90	1.42	10.88
cut out defective 25mm thick tongued and grooved boarding and renew	m2	1.38	11.04	12.01	3.46	26.51
Cut out defective skirting and fit new						
75mm high	m	0.70	5.60	0.68	0.94	7.22
100mm high	m	0.85	6.80	0.96	1.16	8.92
150mm high	m	1.05	8.40	2.05	1.57	12.02
Ease and adjust, oil hardware and ironmongery						
door	nr	1.00	8.00	0.00	1.20	9.20
casement window	nr	0.75	6.00	0.00	0.90	6.90
ease and adjust, oil hardware ironmongery, renew cords to sash window	nr	2.00	16.00	2.31	2.75	21.06

Structural steelwork

Labour rate 8.00 per hour

Remove the following (no allowance made for credits)

Beams, lintels, columns or stanchions, size						
127 x 76mm	m	0.50	4.00	0.00	0.60	4.60
152 x 76mm	m	0.75	6.00	0.00	0.90	6.90

	Unit	Labour hours	Net labour (£)	Net material (£)	O'heads /profit (£)	Total (£)
152 x 152mm	m	1.00	8.00	0.00	1.20	9.20
203 x 142mm	m	1.30	10.40	0.00	1.56	11.96
203 x 203mm	m	1.80	14.40	0.00	2.16	16.56
254 x 114mm	m	1.75	14.00	0.00	2.10	16.10
254 x 146mm	m	2.00	16.00	0.00	2.40	18.40
305 x 165mm	m	3.50	28.00	0.00	4.20	32.20
disconnecting ends of adjacent members	nr	1.50	12.00	0.00	1.80	13.80

Plumbing work

Labour rate 9.60 per hour

Remove 1.83m length of existing gutter, prepare ends and install new length of gutter to existing brackets

Cast iron, half round, diameter

100mm	nr	2.00	19.20	12.53	4.76	36.49
115mm	nr	2.10	20.16	13.45	5.04	38.65
125mm	nr	2.15	20.64	15.30	5.39	41.33

Remove existing gutter fittings, prepare ends and install new fitting

Cast iron, half round, 115mm

angle	nr	1.00	9.60	5.24	2.23	17.07
outlet	nr	1.00	9.60	5.24	2.23	17.07
stop end	nr	0.50	4.80	2.24	1.06	8.10

Remove existing brackets from gutters and replace with galvanized steel repair brackets at 1m maximum centres

Cast iron, half round, diameter

100mm	m	0.25	2.40	1.59	0.60	4.59
112mm	m	0.27	2.59	1.59	0.63	4.81
125mm	m	0.30	2.88	1.73	0.69	5.30

RATES FOR MEASURED WORK

Repairs (cont'd)	Unit	Labour hours	Net labour (£)	Net material (£)	O'heads /profit (£)	Total (£)
Remove 1.83m length of existing pipe, replace with new						
Cast iron, diameter						
65mm	nr	1.80	17.28	14.65	4.79	36.72
75mm	nr	1.90	18.24	24.65	6.43	49.32
100mm	nr	2.05	19.68	33.07	7.91	60.66
Remove existing pipe fittings, replace with new						
Cast iron shoes eared						
65mm	nr	0.66	6.34	11.62	2.69	20.65
75mm	nr	0.80	7.68	11.62	2.89	22.19
100mm	nr	0.90	8.64	15.11	3.56	27.31
Cast iron bends						
65mm	nr	0.75	7.20	7.13	2.15	16.48
75mm	nr	0.95	9.12	8.24	2.60	19.96
100mm	nr	1.00	9.60	11.63	3.18	24.41
Cut out 500mm length of pipe, install new length of pipe with new brass compression connection to existing ends						
Copper						
15mm	nr	0.67	6.43	3.28	1.46	11.17
22mm	nr	0.69	6.62	5.21	1.77	13.60
28mm	nr	0.83	7.97	9.47	2.62	20.06
35mm	nr	0.96	9.22	18.48	4.16	31.86
42mm	nr	1.01	9.70	23.56	4.99	38.25
54mm	nr	1.12	10.75	34.40	6.77	51.92
Take off existing radiator valve replace with new, drain down system prior to removal and bleed after installation						
Single valves						
standard type	nr	1.50	14.40	5.01	2.91	22.32
Complete systems, 9nr valves						
standard type	nr	4.60	44.16	45.08	13.39	102.63

ALTERATIONS - SPOT ITEMS

	Unit	Labour hours	Net labour (£)	Net material (£)	O'heads /profit (£)	Total (£)
Take out existing galvanized steel water storage tank, install new plastic tank complete with ball valve, lid and insulation, allow for cutting holes, tank connectors, make up pipework and connectors to existing pipework						
15 gallon tank	nr	4.00	38.40	35.13	11.03	84.56
25 gallon tank	nr	4.50	43.20	47.64	13.63	104.47
40 gallon tank	nr	5.50	52.80	79.92	19.91	152.63
Take off and relocate existing radiator, drain down system prior to removal, bleed after installation, allow for new surface mounted pipework and removing and refixing existing brackets, compresssion connections to existing pipe						
Distance of relocation						
not exceeding 2m	nr	2.50	24.00	9.56	5.03	38.59
2 to 3m	nr	2.75	26.40	11.36	5.66	43.42
3 to 4m	nr	3.00	28.80	12.72	6.23	47.75
4 to 5m	nr	3.25	31.20	15.96	7.07	54.23

Sanitary fittings

	Unit	Labour hours	Net labour (£)	Net material (£)	O'heads /profit (£)	Total (£)
Take out cast iron bath complete, cut back pipework and prepare to receive new, install new plastic bath complete with trap with integral overflow, pair of pillar taps, waste and plug, make up pipework and connections to hot and cold water and waste pipework						
new bath size 1700 x 700mm	nr	8.00	76.80	190.32	40.07	307.19

Repairs (cont'd)	Unit	Labour hours	Net labour (£)	Net material (£)	O'heads /profit (£)	Total (£)
Take out high level WC and connecting pipework including overflow, cut off spigot from cast iron soil pipe, install new low level WC suite complete including new overflow, Quickfit connector to soil pipe and connection to cold water supply						
new WC suite, size 780 x 510 x 710mm	nr	4.50	43.20	175.86	5.48	224.54

Glazing

Labour rate 8.00 per hour

Hack out glass and remove	m2	0.40	3.20	0.00	0.48	3.68
Clean rebates, remove sprigs or clips and prepare for reglazing	m	0.20	1.60	0.00	0.24	1.84

Paperhanging

Labour rate 8.00 per hour

Strip off one layer of paper, stop cracks, rub down

Woodchip						
walls	m2	0.18	1.44	0.03	0.22	1.69
walls in staircase areas	m2	0.20	1.60	0.03	0.24	1.87
ceilings	m2	0.22	1.76	0.03	0.27	2.06
ceilings in staircase areas	m2	0.24	1.92	0.03	0.29	2.24
Vinyl						
walls	m2	0.23	1.84	0.03	0.28	2.15
walls in staircase areas	m2	0.25	2.00	0.03	0.30	2.33
ceilings	m2	0.27	2.16	0.03	0.33	2.52
ceilings in staircase areas	m2	0.29	2.32	0.03	0.35	2.70
Standard patterned						
walls	m2	0.21	1.68	0.03	0.26	1.97
walls in staircase areas	m2	0.23	1.84	0.03	0.28	2.15

	Unit	Labour hours	Net labour (£)	Net material (£)	O'heads /profit (£)	Total (£)
ceilings	m2	0. 25	2. 00	0. 03	0. 30	2. 33
ceilings in staircase areas	m2	0. 27	2. 16	0. 03	0. 33	2. 52

Painting

Labour rate 8.00 per hour

Preparing existing surfaces
internally

Wash down water painted surfaces
over 300mm girth, stop cracks,
rub down

brickwork walls	m2	0. 14	1. 12	0. 03	0. 17	1. 32
brickwork walls in staircase areas	m2	0. 16	1. 28	0. 03	0. 20	1. 51
blockwork walls	m2	0. 15	1. 20	0. 03	0. 18	1. 41
blockwork walls in staircase areas	m2	0. 17	1. 36	0. 03	0. 21	1. 60

Wash down previously oil painted
wood surfaces, rub down general
surfaces

over 300mm girth	m2	0. 28	2. 24	0. 00	0. 34	2. 58
isolated surfaces not exceeding 300mm girth	m	0. 10	0. 80	0. 00	0. 12	0. 92
isolated areas not exceeding 0.5m2	nr	0. 12	0. 96	0. 00	0. 14	1. 10

	Unit	Labour hours	Net labour (£)	Net material (£)	O'heads /profit (£)	Total (£)

C40 REPAIR/RENOVATE MASONRY

Labour rate 8.00 per hour

Make good areas of concrete after removal of parts

Face of concrete wall

250 to 500mm wide	m	1.00	8.00	0.00	0.20	8.20
500 to 1000mm wide	m	1.10	8.80	0.00	0.22	9.02
over 1000mm wide	m2	1.20	9.60	0.00	1.44	11.04

Dress edge of concrete slab to receive new concrete

150mm thick	m	0.10	0.80	0.00	0.12	0.92
200mm thick	m	0.20	1.60	0.00	0.24	1.84
250mm thick	m	0.25	2.00	0.00	0.30	2.30

Make good surface of reinforced concrete floor slab after removal of column, prepare for new screed or floor finish

area not exceeding 1m2	nr	0.30	2.40	0.00	0.06	2.46
area 1m2 to 2m2	nr	0.50	4.00	0.00	0.10	4.10

Make good exposed face of existing concrete after cutting back

up to 200mm wide	m	0.35	2.80	0.00	0.42	3.22
200 to 500mm wide	m	0.60	4.80	0.00	0.72	5.52
500 to 1000mm wide	m	1.20	9.60	0.00	1.44	11.04
over 1000mm wide	m2	1.20	9.60	0.00	1.44	11.04
prepare, clean face of existing concrete to receive new finish	m2	0.90	7.20	0.00	1.08	8.28
prepare face of existing brickwork, rake out joints to receive cement render or plaster	m2	0.36	2.88	0.00	0.43	3.31

REPAIRING/RENOVATING BRICKWORK

	Unit	Labour hours	Net labour (£)	Net material (£)	O'heads /profit (£)	Total (£)
Rake out joints of existing brickwork, point in cement mortar (1:3)						
flush pointing	m2	1. 11	8. 88	0. 32	1. 38	10. 58
weather pointing	m2	1. 16	9. 28	0. 32	1. 44	11. 04
Cut out single brick in half brick walls, replace in gauged mortar (1:1:6)						
commons	nr	0. 26	2. 08	0. 16	0. 34	2. 58
facings	nr	0. 30	2. 40	0. 46	0. 43	3. 29
Cut out single brick in one brick wall, replace in gauged mortar (1:1:6)						
commons	nr	0. 38	3. 04	0. 16	0. 48	3. 68
facings	nr	0. 44	3. 52	0. 46	0. 60	4. 58
Cut out single brick in half brick wall, replace in cement mortar (1:3)						
commons	nr	0. 26	2. 08	0. 16	0. 34	2. 58
facings	nr	0. 30	2. 40	0. 46	0. 43	3. 29
Cut out single brick in one brick wall, replace in cement mortar (1:3)						
commons	nr	0. 38	3. 04	0. 16	0. 48	3. 68
facings	nr	0. 40	3. 20	0. 45	0. 55	4. 20
Cut out decayed brickwork in half brick wall, areas 0.5 to 1m2, replace in gauged mortar (1:1:6)						
commons	nr	5. 20	41. 60	10. 40	7. 80	59. 80
facings	nr	6. 00	48. 00	28. 55	11. 48	88. 03

Cutting out (cont'd)	Unit	Labour hours	Net labour (£)	Net material (£)	O'heads /profit (£)	Total (£)
Cut out decayed brickwork in one brick wall, areas 0.5 to 1m2, replace in gauged mortar (1:1:6)						
commons	nr	7.60	60.80	21.21	12.30	94.31
facings	nr	8.80	70.40	53.42	18.57	142.39
Cut out decayed brickwork in half brick wall, areas 0.5 to 1m2, replace in cement mortar (1:3)						
commons	nr	6.28	50.24	9.55	8.97	68.76
facings	nr	7.24	57.92	26.98	12.73	97.63
Cut out decayed brickwork in one brick wall, areas 0.5 to 1m2, replace in cement mortar (1:3)						
commons	nr	9.38	75.04	21.70	14.51	111.25
facings	nr	10.64	85.12	53.92	20.86	159.90
Cut out vertical, horizontal or stepped cracks in half brick wall, replace average 350mm width in gauged mortar (1:1:6)						
commons	nr	2.76	22.08	3.76	3.88	29.72
facings	nr	3.12	24.96	9.78	5.21	39.95
Cut out vertical, horizontal or stepped cracks in one brick wall, replace average 350mm width in gauged mortar (1:1:6)						
commons	nr	5.18	41.44	9.88	7.70	59.02
facings	nr	6.10	48.80	19.52	10.25	78.57
Cut out vertical, horizontal or stepped cracks in one and a half brick wall, replace average 350mm width in gauged mortar (1:1:6)						
commons	nr	8.12	64.96	11.67	11.49	88.12
facings	nr	9.25	74.00	29.54	15.53	119.07

REPAIRING/RENOVATING BRICKWORK

	Unit	Labour hours	Net labour (£)	Net material (£)	O'heads /profit (£)	Total (£)
Cut out vertical, horizontal or stepped cracks in half brick wall, replace average 350mm width in cement mortar (1:3)						
commons	m	3.38	27.04	3.84	4.63	35.51
facings	m	3.50	28.00	9.87	5.68	43.55
Cut out vertical, horizontal or stepped cracks in one brick wall, replace average 350mm width in cement mortar (1:3)						
commons	m	6.20	49.60	7.88	8.62	66.10
facings	m	7.22	57.76	19.71	11.62	89.09
Cut out vertical, horizontal or stepped cracks in one and a half brick wall, replace average 350mm width in cement mortar (1:3)						
commons	m	9.72	77.76	11.98	13.46	103.20
facings	m	11.10	88.80	20.84	16.45	126.09
Cut out defective brick on end soldier arch to half brick thick, replace in gauged mortar (1:1:6)						
commons	m	0.81	6.48	2.08	1.28	9.84
facings	m	0.94	7.52	5.71	1.98	15.21
Cut out defective brick on end soldier arch to half brick thick, replace in cement mortar (1:3)						
commons	m	0.98	7.84	2.08	1.49	11.41
facings	m	1.14	9.12	5.40	2.18	16.70
Cut out defective segmental arch half brick thick, replace in gauged mortar (1:1:6)						
commons	m	1.24	9.92	2.08	1.80	13.80
facings	m	1.38	11.04	5.71	2.51	19.26

Cutting out (cont'd)	Unit	Labour hours	Net labour (£)	Net material (£)	O'heads /profit (£)	Total (£)
Cut out defective segmental arch one brick thick, replace in gauged mortar (1:1:6)						
commons	m	1.41	11.28	2.08	2.00	15.36
facings	m	1.65	13.20	5.71	2.84	21.75
Cut out defective brick on edge arch one brick thick, replace in gauged mortar (1:1:6)						
commons	m	0.84	6.72	4.24	1.64	12.60
facings	m	1.05	8.40	10.63	2.85	21.88
Cut out defective segmental arch half brick thick, replace in cement mortar (1:3)						
commons	m	1.50	12.00	2.08	2.11	16.19
facings	m	1.57	12.56	5.71	2.74	21.01
Cut out defective segmental arch one brick thick, replace in cement mortar (1:3)						
commons	m	1.69	13.52	2.08	2.34	17.94
facings	m	1.98	15.84	5.71	3.23	24.78
Cut out defective brick on edge arch one brick thick, replace in cement mortar (1:3)						
commons	m	1.02	8.16	4.34	1.88	14.38
facings	m	1.29	10.32	10.78	3.17	24.27
Cut out defective terracotta air brick, replace						
215 x 65mm	nr	0.35	2.80	1.85	0.70	5.35
215 x 140mm	nr	0.50	4.00	2.82	1.02	7.84

REPAIRING/RENOVATING BRICKWORK

	Unit	Labour hours	Net labour (£)	Net material (£)	O'heads /profit (£)	Total (£)
Rake out joints in cement mortar (1:3), refix existing flashing, point up						
horizontal	m	0.38	3.04	0.32	0.50	3.86
stepped	m	0.54	4.32	0.44	0.71	5.47
Rake out joints in existing brickwork and point up in cement mortar (1:3)	m2	0.90	7.20	0.52	1.16	8.88
Cut out one course of half brick wall, insert hessian based damp-proof course 112mm wide, replace with new bricks in gauged mortar (1:1:6)						
commons	m	1.92	15.36	1.28	2.50	19.14
facings	m	1.92	15.36	2.36	2.66	20.38
Cut out one course of one brick wall, insert hessian based damp-proof course 225mm wide, replace with new bricks in gauged mortar (1:1:6)						
commons	m	4.20	33.60	2.57	5.43	41.60
facings	m	4.20	33.60	4.48	5.71	43.79
Cut out one course of half brick wall, insert hessian based damp-proof course 112mm wide, replace with new bricks in cement mortar (1:3)						
commons	m	2.28	18.24	1.22	2.92	22.38
facings	m	2.28	18.24	1.95	3.03	23.22

Cutting out (cont'd)	Unit	Labour hours	Net labour (£)	Net material (£)	O'heads /profit (£)	Total (£)
Cut out one course of one brick wall, insert hessian based damp-proof course 225mm wide, replace with new bricks in cement mortar (1:3)						
commons	m	4. 86	38. 88	1. 85	6. 11	46. 84
facings	m	4. 86	38. 88	4. 54	6. 51	49. 93
Cut out one course of half brick wall, insert pitch polymer damp-proof course 112mm wide, replace with new bricks in gauged mortar (1:1:6)						
commons	m	1. 92	15. 36	1. 36	2. 51	19. 23
facings	m	1. 92	15. 36	2. 44	2. 67	20. 47
Cut out one course of one brick wall, insert pitch polymer damp-proof course 225mm wide, replace with new bricks in gauged mortar (1:1:6)						
commons	m	4. 20	33. 60	2. 72	5. 45	41. 77
facings	m	4. 20	33. 60	4. 63	5. 73	43. 96
Cut out one course of half brick wall, insert pitch polymer damp-proof course 112mm wide, replace with new bricks in cement mortar (1:3)						
commons	m	2. 28	18. 24	1. 30	2. 93	22. 47
facings	m	2. 28	18. 24	2. 03	3. 04	23. 31

	Unit	Labour hours	Net labour (£)	Net material (£)	O'heads /profit (£)	Total (£)
Cut out one course of one brick wall, insert pitch polymer damp-proof course 225mm wide, replace with new bricks in cement mortar (1:3)						
commons	m	4.86	38.88	3.07	6.29	48.24
facings	m	4.86	38.88	4.68	6.53	50.09
Cut out one course of half brick wall, insert bitumen based lead cored damp-proof course 112mm wide, replace with new bricks in gauged mortar (1:1:6)						
commons	m	1.92	15.36	2.21	2.64	20.21
facings	m	1.92	15.36	3.29	2.80	21.45
Cut out one course of one brick wall, insert bitumen based lead cored damp-proof course 225mm wide, replace with new bricks in gauged mortar (1:1:6)						
commons	m	4.20	33.60	4.33	5.69	43.62
facings	m	4.20	33.60	6.34	5.99	45.93
Cut out one course of half brick wall, insert bitumen based lead cored damp-proof course 112mm wide, replace with new bricks in cement mortar (1:3)						
commons	m	2.28	18.24	2.15	3.06	23.45
facings	m	2.28	18.24	2.88	3.17	24.29

Cutting out (cont'd)	Unit	Labour hours	Net labour (£)	Net material (£)	O'heads /profit (£)	Total (£)
Cut out one course of one brick wall, insert bitumen based lead cored damp-proof course 225mm wide, replace with new bricks in cement mortar (1:3)						
commons	m	4. 86	38. 88	4. 77	6. 55	50. 20
facings	m	4. 86	38. 88	6. 40	6. 79	52. 07
Cut out single block, replace in cement mortar (1:3) thickness						
100mm	nr	0. 28	2. 24	0. 77	0. 45	3. 46
150mm	nr	0. 42	3. 36	1. 04	0. 66	5. 06
200mm	nr	0. 50	4. 00	1. 54	0. 83	6. 37
250mm	nr	0. 56	4. 48	2. 24	1. 01	7. 73
Cut out areas not exceeding 1m2 in random rubble walling, clean off and reset in gauged mortar (1:1:6), thickness						
200mm	m2	4. 30	34. 40	0. 91	5. 30	40. 61
300mm	m2	5. 20	41. 60	1. 38	6. 45	49. 43
450mm	m2	6. 85	54. 80	2. 04	8. 53	65. 37
Cut out areas not exceeding 1m2 in coursed rubble walling, clean off and reset in gauged mortar (1:1:6), thickness						
200mm	m2	4. 50	36. 00	0. 91	5. 54	42. 45
300mm	m2	5. 20	41. 60	1. 38	6. 45	49. 43
450mm	m2	7. 05	56. 40	2. 04	8. 77	67. 21
Rake out joints of rubble walling, point in gauged mortar (1:1:6)						
Uncoursed						
flush pointing	m2	0. 48	3. 84	0. 32	0. 62	4. 78
weather pointing	m2	0. 48	3. 84	0. 32	0. 62	4. 78

REPAIRING/RENOVATING BRICKWORK

	Unit	Labour hours	Net labour (£)	Net material (£)	O'heads /profit (£)	Total (£)
Coursed						
flush pointing	m2	0.48	3.84	0.32	0.62	4.78
weather pointing	m2	0.48	3.84	0.32	0.62	4.78

D

GROUND WORK

D20 Excavation and filling

MATERIAL COSTS

	Unit	Material supply (£)	Material waste (%)	Total (£)
MATERIALS				
Bulk materials				
Sand	m3	15. 50	10. 0	17. 05
Hardcore	m3	9. 30	10. 0	10. 23

RATES FOR MEASURED WORK

	Unit	Labour hours	Net labour (£)	Net material (£)	O'heads /profit (£)	Total (£)

D20 EXCAVATION AND FILLING

Labour rate 7.00 per hour

HAND EXCAVATION

The following unit rates refer to
work carried out in firm ground.
The following adjustments should
be applied for other ground
conditions.

Clay   +75%

Soft chalk +125%

Cut down trees, grub up roots

	Unit	Labour hours	Net labour (£)	Net material (£)	O'heads /profit (£)	Total (£)
600mm to 1.5m girth	nr	18.10	126.70	0.00	19.00	145.70
1.5 to 3.0m girth	nr	30.00	210.00	0.00	31.50	241.50

Cut down hawthorn hedge, grub up
roots

	Unit	Labour hours	Net labour (£)	Net material (£)	O'heads /profit (£)	Total (£)
1500mm high	m	2.10	14.70	0.00	2.20	16.90
3000mm high	m	2.30	16.10	0.00	2.42	18.52

Excavate topsoil, lay aside for
reuse, average depth

	Unit	Labour hours	Net labour (£)	Net material (£)	O'heads /profit (£)	Total (£)
150mm	m2	0.37	2.59	0.00	0.39	2.98
200mm	m2	0.47	3.29	0.00	0.49	3.78

Excavate from ground level to
reduce levels, maximum depth not
exceeding

	Unit	Labour hours	Net labour (£)	Net material (£)	O'heads /profit (£)	Total (£)
0.25m	m3	2.50	17.50	0.00	2.62	20.12
1.00m	m3	2.75	19.25	0.00	2.89	22.14
2.00m	m3	3.00	21.00	0.00	3.15	24.15

EXCAVATION AND FILLING

	Unit	Labour hours	Net labour (£)	Net material (£)	O'heads /profit (£)	Total (£)
Excavate pits to receive bases, maximum depth not exceeding						
0.25m	m3	3.25	22.75	0.00	3.41	26.16
1.00m	m3	4.00	28.00	0.00	4.20	32.20
2.00m	m3	4.50	31.50	0.00	4.72	36.22
Excavate trenches to receive foundations, maximum depth not exceeding						
0.25m	m3	2.70	18.90	0.00	2.83	21.73
1.00m	m3	3.00	21.00	0.00	3.15	24.15
2.00m	m3	3.75	26.25	0.00	3.94	30.19
Excavate trenches for services not exceeding 300mm wide, back fill with excavated material						
Extra for breaking up						
concrete 100mm thick	m2	0.90	6.30	0.00	0.94	7.24
tarmacadam 75mm thick	m2	0.50	3.50	0.00	0.53	4.03
hardcore 100mm thick	m2	0.60	4.20	0.00	0.63	4.83
plain concrete	m2	7.00	49.00	0.00	7.35	56.35
reinforced concrete	m2	8.00	56.00	0.00	8.40	64.40
soft rock	m2	10.00	70.00	0.00	10.50	80.50
hard rock	m2	11.00	77.00	0.00	11.55	88.55

MACHINE EXCAVATION

The following entered under the Net material column refer to plant costs.

The following unit rates refer to work carried out in firm ground. The following adjustments should be applied for other ground conditions.

Clay +25%

Soft chalk +50%

RATES FOR MEASURED WORK

Machine excavation (cont'd)	Unit	Labour hours	Net labour (£)	Net material (£)	Net plant (£)	O'heads /profit (£)	Total (£)
Cut down trees, grub up roots							
600mm to 1.5m girth	nr	7.00	49.00	0.00	43.00	13.80	105.80
1.5 to 3.0m girth	nr	8.00	56.00	0.00	78.00	20.10	154.10
Cut down hawthorn hedge, grub up roots							
1500mm high	m	0.50	3.50	0.00	8.00	1.72	13.23
3000mm high	m	0.60	4.20	0.00	8.00	1.83	14.03
Excavate topsoil, lay aside for reuse, average depth							
150mm	m2	0.02	0.14	0.00	0.40	0.08	0.62
200mm	m2	0.03	0.21	0.00	0.45	0.10	0.76
Excavate from ground level to reduce levels, maximum depth not exceeding							
0.25m	m3	0.10	0.70	0.00	1.30	0.30	2.30
1.00m	m3	0.10	0.70	0.00	1.15	0.28	2.13
2.00m	m3	0.10	0.70	0.00	1.40	0.32	2.42
Excavate pits to receive bases, maximum depth not exceeding							
0.25m	m3	0.30	2.10	0.00	5.50	1.14	8.74
1.00m	m3	0.25	1.75	0.00	6.00	1.16	8.91
2.00m	m3	0.28	1.96	0.00	6.50	1.27	9.73
Excavate trenches to receive foundations, maximum depth not exceeding							
0.25m	m3	0.35	2.45	0.00	5.50	1.19	9.14
1.00m	m3	0.30	2.10	0.00	6.00	1.21	9.32
2.00m	m3	0.34	2.38	0.00	6.50	1.33	10.21

EXCAVATION AND FILLING

	Unit	Labour hours	Net labour (£)	Net material (£)	Net plant (£)	O'heads /profit (£)	Total (£)
Excavate trenches for services not exceeding 300mm wide, backfill with excavated material							
Extra for breaking up by compressor and tools							
concrete 100mm thick	m2	0.40	2.80	0.00	2.50	0.79	6.10
tarmacadam 75mm thick	m2	0.24	1.68	0.00	1.30	0.45	3.43
hardcore 100mm thick	m2	0.28	1.96	0.00	1.40	0.50	3.86
plain concrete	m2	0.30	2.10	0.00	16.00	2.72	20.82
reinforced concrete	m2	3.50	24.50	0.00	22.00	6.97	53.48
soft rock	m2	4.00	28.00	0.00	14.00	6.30	48.30
hard rock	m2	4.40	30.80	0.00	22.00	7.92	60.72

Earthwork support

	Unit	Labour hours	Net labour (£)	Net material (£)	Net plant (£)	O'heads /profit (£)	Total (£)
Earthwork support not exceeding 2m between opposing faces, depth not exceeding 1m in							
firm ground	m2	0.40	2.80	1.10	0.00	0.58	4.48
loose ground	m2	1.25	8.75	1.87	0.00	1.59	12.21
sand	m2	2.75	19.25	3.08	0.00	3.35	25.68
Earthwork support not exceeding 2m between opposing faces, depth not exceeding 2m in							
firm ground	m2	0.65	5.20	1.21	0.00	0.96	7.37
loose ground	m2	1.35	9.45	2.09	0.00	1.73	13.27
sand	m2	3.15	22.05	3.19	0.00	3.79	29.03
Earthwork support not exceeding 2m between opposing faces, depth not exceeding 4m in							
firm ground	m2	0.65	4.55	1.32	0.00	0.88	6.75
loose ground	m2	1.50	10.50	2.42	0.00	1.94	14.86
sand	m2	3.50	24.50	4.40	0.00	4.33	33.23

RATES FOR MEASURED WORK

Machine excavation (cont'd)	Unit	Labour hours	Net labour (£)	Net material (£)	O'heads /profit (£)	Total (£)
Disposal						
Load surplus excavated material into barrows, wheel and deposit in temporary spoil heaps, average distance						
25m	m3	1.20	8.40	0.00	1.26	9.66
50m	m3	2.00	14.00	0.00	2.10	16.10
Load into barrows, wheel and deposit in skip or lorry, average distance						
25m	m3	1.20	8.40	0.00	1.26	9.66
Remove from site by skip (5m3) including tipping charges	m3	0.00	0.00	10.78	1.62	12.40
Remove from site by lorry (16 tonne) including tipping charges (average 10km to tip)	m3	0.00	0.00	14.30	0.36	14.66
Filling						
Filling material deposited on site in layers not exceeding 250mm thick, average distance 25m, compacting with vibrating roller						
surplus excavated material	m3	0.33	2.31	4.20	0.98	7.49
sand	m3	0.42	2.94	17.05	3.00	22.99
hardcore	m3	0.42	2.94	10.23	1.98	15.15
Surface treatments						
Level and compact bottom of excavation with vibrating roller						
surplus excavated material	m2	0.12	0.84	0.66	0.22	1.72
sand	m2	0.11	0.77	0.61	0.21	1.59
hardcore	m2	0.12	0.84	0.66	0.22	1.72

E
IN SITU CONCRETE

Estimate_It

The easy to use computer estimating program for contractors, surveyors, and estimators. Incorporating data from the Spons's Contractors Handbooks series, prices can be adjusted to reflect local conditons.

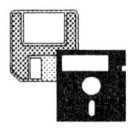

Seven libraries available

Minor Works
Plumbing & Domestic Heating
Painting, Decorating & Glazing
Electrical Installation
Roofing
Floor, Wall & Ceiling Finishes
Landscaping

The Measuring Mouse Measure direct from drawing quickly with our measuring mouse.

Composite Rates Build your own composites for quick estimating.

Superb Reports Full detailed break down of labour, material, and plant costs, with presentation you can be proud of.

Post Contract Track projects for progress and scope changes.

For further information please complete and return this page (or photocopy) to Snape.

Name _____

Address _____

_____

Post Code _____ Tel _____ Fax _____

Snape Computers Ltd. 134 High Street, Ponders End, Enfield EN3 4ET
Tel 081 805 8704   Fax 081 443 5828

Snape

MATERIAL COSTS

	Unit	Material supply (£)	Material waste (%)	Total (£)
MATERIAL COSTS				
Sand	m3	15.50	10.0	17.05
Cement	t	87.75	10.0	96.53
Sulphate resisting cement	t	92.21	10.0	101.43
Aggregate				
20mm	t	11.50	10.0	12.65
40mm	t	11.30	10.0	12.43
Site mixed concrete				
(Mix A2-1:3:6)	m3	48.59	5.0	51.02
(Mix B2-1:2:4)	m3	56.44	5.0	59.26
Ready mixed concrete (full loads)				
(Mix A-1:3:6)	m3	46.16	5.0	48.47
(Mix B-1:2:4)	m3	53.62	5.0	56.30
Bituminous emulsion waterproofer	5l	9.50	5.0	9.98
Plain round reinforcement bars BS4449				
6mm	m	0.11	5.0	0.12
8mm	m	0.20	5.0	0.21
10mm	m	0.31	5.0	0.33
12mm	m	0.45	5.0	0.47
16mm	m	0.88	5.0	0.92
20mm	m	1.30	5.0	1.37
25mm	m	2.01	5.0	2.11

**Steel sheet fabric reinforcement
in standard sheets to BS4483**

Ref	kg/m2				
A98	1.54	m2	1.17	5.0	1.23
A142	2.22	m2	1.54	5.0	1.62
A193	3.02	m2	2.10	5.0	2.21
A252	3.95	m2	2.75	5.0	2.89
A393	6.16	m2	4.29	5.0	4.50

Steel sheet fabric (cont'd)		Unit	Material supply (£)	Material waste (%)	Total (£)
B196	3.05	m2	2.22	5.0	2.33
B283	3.73	m2	2.67	5.0	2.80
B385	4.53	m2	3.21	5.0	3.37
B503	5.93	m2	4.18	5.0	4.39
B785	8.14	m2	5.73	5.0	6.02
B1131	10.90	m2	7.68	5.0	8.06
C283	2.61	m2	1.90	5.0	2.00
C385	3.41	m2	2.45	5.0	2.57
C503	4.34	m2	3.07	5.0	3.22
C636	5.55	m2	3.96	5.0	4.16
C785	6.72	m2	4.80	5.0	5.04
Sawn softwood, 47 x 75mm		m	1.17	5.0	1.23
12mm softwood		m2	10.60	5.0	11.13

Precast prestressed concrete lintels

65 x 100mm

	Unit	Material supply (£)	Material waste (%)	Total (£)
1.35m long	nr	4.91	5.0	5.16
2.25m long	nr	8.18	5.0	8.59

65 x 150mm

	Unit	Material supply (£)	Material waste (%)	Total (£)
1.35m long	nr	6.31	5.0	6.63
2.25m long	nr	10.53	5.0	11.06

65 x 220mm

	Unit	Material supply (£)	Material waste (%)	Total (£)
1.35m long	nr	10.29	5.0	10.80
2.25m long	nr	17.15	5.0	18.01

Precast concrete copings once weathered and throated

	Unit	Material supply (£)	Material waste (%)	Total (£)
75 x 300 x 900mm	nr	5.90	5.0	6.20
75 x 300 x 600mm	nr	4.10	5.0	4.31

MATERIAL COSTS

	Unit	Material supply (£)	Material waste (%)	Total (£)
Precast concrete saddleback copings twice weathered and throated				
75 x 300 x 600mm	nr	4.10	5.0	4.31
60 x 160 x 450mm	nr	2.70	5.0	2.84
Flexcell joint filler 12.5mm thick, width over				
150mm	m	1.85	5.0	1.94
300mm	m	2.69	5.0	2.82
450mm	m	7.06	5.0	7.41
Flexcell joint filler 20mm thick, width over				
150mm	m	2.53	5.0	2.66
300mm	m	3.68	5.0	3.86
450mm	m	9.18	5.0	9.64

RATES FOR MEASURED WORK

	Unit	Labour hours	Net labour (£)	Net material (£)	O'heads /profit (£)	Total (£)
E10 IN SITU CONCRETE						
Labour rate 8.00 per hour						
SITE MIXED CONCRETE: MIX A						
Foundations in trenches	m3	1.35	10.80	51.02	9.27	71.09
Isolated bases	m3	1.90	15.20	51.02	9.93	76.15
Filling to hollow walls not exceeding 150mm thick	m3	4.10	32.80	51.02	12.57	96.39
Extra for placing concrete around reinforcement	m3	0.55	4.40	0.00	0.66	5.06
SITE MIXED CONCRETE: MIX B						
Beds						
over 450mm thick	m3	1.35	10.80	59.26	10.51	80.57
150 to 450mm thick	m3	1.70	13.60	59.26	10.93	83.79
not exceeding 150mm thick	m3	2.80	22.40	59.26	12.25	93.91
Suspended slabs						
over 450mm thick	m3	1.35	10.80	59.26	10.51	80.57
150 to 450mm thick	m3	1.70	13.60	59.26	10.93	83.79
not exceeding 150mm thick	m3	2.80	22.40	59.26	12.25	93.91
Walls						
over 450mm thick	m3	2.60	20.80	59.26	12.01	92.07
150 to 450mm thick	m3	3.60	28.80	59.26	13.21	101.27
not exceeding 150mm thick	m3	4.20	33.60	59.26	13.93	106.79
Casings						
casings to isolated beams	m3	5.00	40.00	59.26	14.89	114.15
casings to isolated deep beam	m3	4.50	36.00	59.26	14.29	109.55
casings to attached deep beam	m3	4.50	36.00	59.26	14.29	109.55
isolated beam	m3	5.00	40.00	59.26	14.89	114.15
isolated deep beam	m3	4.50	36.00	59.26	14.29	109.55
attached deep beam	m3	4.50	36.00	59.26	14.29	109.55

IN SITU CONCRETE

	Unit	Labour hours	Net labour (£)	Net material (£)	O'heads /profit (£)	Total (£)
isolated beam casings	m3	4.50	36.00	59.26	14.29	109.55
isolated deep beam casings	m3	4.50	36.00	59.26	14.29	109.55
attached deep beam casings	m3	4.50	36.00	59.26	14.29	109.55
columns	m3	5.00	40.00	59.26	14.89	114.15
column casings	m3	5.00	40.00	59.26	14.89	114.15
staircases	m3	6.00	48.00	59.26	16.09	123.35

READY MIXED CONCRETE: MIX A

	Unit	Labour hours	Net labour (£)	Net material (£)	O'heads /profit (£)	Total (£)
Foundations in trenches	m3	1.35	10.80	48.47	8.89	68.16
Isolated bases	m3	1.90	15.20	48.47	9.55	73.22
Filling to hollow walls not exceeding 150mm thick	m3	4.10	32.80	48.47	12.19	93.46
Extra for placing concrete around reinforcement	m3	0.55	4.40	0.00	0.66	5.06

READY MIXED CONCRETE: MIX B

Beds

	Unit	Labour hours	Net labour (£)	Net material (£)	O'heads /profit (£)	Total (£)
over 450mm thick	m3	1.35	10.80	56.30	10.06	77.16
150 to 450mm thick	m3	1.70	13.60	56.30	10.48	80.38
not exceeding 150mm thick	m3	2.80	22.40	56.30	11.80	90.50

Suspended slabs

	Unit	Labour hours	Net labour (£)	Net material (£)	O'heads /profit (£)	Total (£)
over 450mm thick	m3	1.35	10.80	56.30	10.06	77.16
150 to 450mm thick	m3	1.70	13.60	56.30	10.48	80.38
not exceeding 150mm thick	m3	2.80	22.40	56.30	11.80	90.50

Walls

	Unit	Labour hours	Net labour (£)	Net material (£)	O'heads /profit (£)	Total (£)
over 450mm thick	m3	2.60	20.80	56.30	11.56	88.66
150 to 450mm thick	m3	3.60	28.80	56.30	12.76	97.86
not exceeding 150mm thick	m3	4.20	33.60	56.30	13.48	103.38

Casings

	Unit	Labour hours	Net labour (£)	Net material (£)	O'heads /profit (£)	Total (£)
casings to isolated beams	m3	5.00	40.00	56.30	14.44	110.74
casings to isolated deep beam	m3	4.50	36.00	56.30	13.84	106.14
casings to attached deep beam	m3	4.50	36.00	56.30	13.84	106.14

Ready mixed concrete (cont'd)	Unit	Labour hours	Net labour (£)	Net material (£)	O'heads /profit (£)	Total (£)
isolated beam	m3	5.00	40.00	56.30	14.44	110.74
isolated deep beam	m3	4.50	36.00	56.30	13.84	106.14
attached deep beam	m3	4.50	36.00	56.30	13.84	106.14
isolated beam casings	m3	5.00	40.00	56.30	14.44	110.74
isolated deep beam casings	m3	4.50	36.00	56.30	13.84	106.14
attached deep beam casings	m3	4.50	36.00	56.30	13.84	106.14
columns	m3	5.00	40.00	56.30	14.44	110.74
column casings	m3	5.00	40.00	56.30	14.44	110.74
staircases	m3	6.00	48.00	56.30	15.64	119.94
Extra for						
laying to slopes not exceeding 15 degrees	m2	0.50	4.00	6.30	1.54	11.84
laying to slopes over 15 degrees	m2	0.70	5.60	0.00	0.84	6.44
sulphate resisting cement	m3	0.00	0.00	12.60	1.89	14.49

FORMWORK FOR IN SITU CONCRETE

	Unit	Labour hours	Net labour (£)	Net material (£)	O'heads /profit (£)	Total (£)
E20 FORMWORK FOR IN SITU CONCRETE						
Labour rate 8.00 per hour						
All the following rates are based on all four uses.						
Sides of foundations, bases, beams or beds, height						
over 1m	m2	2.30	18.40	3.98	3.36	25.74
not exceeding 250mm	m	0.75	6.00	0.99	1.05	8.04
250 to 500mm	m	1.28	10.24	1.99	1.83	14.06
500mm to 1m	m	1.68	13.44	3.70	2.57	19.71
Sides of ground beams and edges of beds						
over 1m	m2	2.30	18.40	3.98	3.36	25.74
not exceeding 250mm	m	0.75	6.00	0.99	1.05	8.04
250 to 500mm	m	1.28	10.24	1.99	1.83	14.06
500mm to 1m	m	1.68	13.44	3.70	2.57	19.71
Edges of suspended slab						
not exceeding 250mm	m	0.90	7.20	0.99	1.23	9.42
250 to 500mm	m	1.40	11.20	1.99	1.98	15.17
500mm to 1m	m	1.80	14.40	3.70	2.71	20.81
Sides of upstands						
over 1m	m2	2.50	20.00	3.98	3.60	27.58
not exceeding 250mm	m	0.90	7.20	0.99	1.23	9.42
250 to 500mm	m	1.40	11.20	1.99	1.98	15.17
500mm to 1m	m	1.80	14.40	3.70	2.71	20.81
Steps in top surfaces						
not exceeding 250mm	m	0.75	6.00	0.99	1.05	8.04
250 to 500mm	m	1.28	10.24	1.99	1.83	14.06
500mm to 1m	m	1.68	13.44	3.70	2.57	19.71

RATES FOR MEASURED WORK

Formwork (cont'd)	Unit	Labour hours	Net labour (£)	Net material (£)	O'heads /profit (£)	Total (£)
Steps in soffits						
not exceeding 250mm	m	0.90	7.20	0.99	1.23	9.42
250 to 500mm	m	1.40	11.20	1.99	1.98	15.17
500mm to 1m	m	1.80	14.40	3.70	2.71	20.81
Machine bases and plinths						
over 1m	m2	2.30	18.40	3.98	3.36	25.74
not exceeding 250mm	m	0.75	6.00	0.99	1.05	8.04
250 to 500mm	m	1.28	10.24	1.99	1.83	14.06
500mm to 1m	m	1.68	13.44	3.70	2.57	19.71
Soffits of slabs						
not exceeding 200mm thick	m2	2.40	19.20	3.98	3.48	26.66
200 to 300mm thick	m2	2.60	20.80	3.98	3.72	28.50
300 to 400mm thick	m2	2.80	22.40	3.98	3.96	30.34
Walls						
vertical	m2	2.45	19.60	3.98	3.54	27.12
vertical, interrupted	m2	2.55	20.40	4.55	3.74	28.69
vertical, exceeding 3m high	m2	2.65	21.20	5.10	3.94	30.24
Beams, rectangular attached to in situ slabs, height to soffit						
1.5 to 3m	m2	3.20	25.60	4.55	4.52	34.67
3 to 4.5m	m2	3.30	26.40	5.10	4.72	36.22
columns, rectangular attached to in situ walls	m2	3.20	25.60	4.55	4.52	34.67
Mortice in concrete for rag bolt, grout in cement mortar (1:1), depth						
50mm	nr	0.25	2.00	0.66	0.40	3.06
100mm	nr	0.30	2.40	1.20	0.54	4.14
150mm	nr	0.35	2.80	1.76	0.68	5.24

FORMWORK FOR IN SITU CONCRETE

	Unit	Labour hours	Net labour (£)	Net material (£)	O'heads /profit (£)	Total (£)
Mortice in concrete for holding down bolt and plates, grout in cement mortar (1:3), depth						
200mm	nr	0.60	4.80	2.30	1.06	8.16
400mm	nr	0.84	6.72	3.45	1.53	11.70

RATES FOR MEASURED WORK

	Unit	Labour hours	Net labour (£)	Net material (£)	O'heads /profit (£)	Total (£)

E30 REINFORCEMENT FOR IN SITU CONCRETE

Labour rate 8.00 per hour

Reinforcement bars, plain round steel to BS4449, straight or bent

	Unit	Labour hours	Net labour (£)	Net material (£)	O'heads/profit (£)	Total (£)
6mm	m	0.12	0.96	0.12	0.16	1.24
8mm	m	0.10	0.80	0.21	0.15	1.16
10mm	m	0.09	0.72	0.33	0.16	1.21
12mm	m	0.08	0.64	0.47	0.17	1.28
16mm	m	0.07	0.56	0.92	0.22	1.70
20mm	m	0.06	0.48	1.37	0.28	2.13
25mm	m	0.05	0.40	2.11	0.38	2.89

Steel fabric reinforcement, BS4483 lapped, laid in concrete beds

Ref   kg/m2

Ref	kg/m2	Unit	Labour hours	Net labour (£)	Net material (£)	O'heads/profit (£)	Total (£)
A98	1.54	m2	0.11	0.88	1.23	0.32	2.43
A142	2.22	m2	0.12	0.96	1.62	0.39	2.97
A193	3.02	m2	0.15	1.20	2.21	0.51	3.92
A252	3.95	m2	0.17	1.36	2.89	0.64	4.89
A393	6.16	m2	0.20	1.60	4.50	0.91	7.01
B196	3.05	m2	0.15	1.20	2.33	0.53	4.06
B283	3.73	m2	0.16	1.28	2.80	0.61	4.69
B385	4.53	m2	0.17	1.36	3.37	0.71	5.44
B503	5.93	m2	0.19	1.52	4.39	0.89	6.80
B785	8.14	m2	0.22	1.76	6.02	1.17	8.95
B1131	10.90	m2	0.24	1.92	8.06	1.50	11.48
C283	2.61	m2	0.13	1.04	2.00	0.46	3.50
C385	3.41	m2	0.16	1.28	2.88	0.62	4.78
C503	4.34	m2	0.18	1.44	3.22	0.70	5.36
C636	5.55	m2	0.19	1.52	4.16	0.85	6.53
C785	6.72	m2	0.21	1.68	5.04	1.01	7.73

Cutting on mesh reinforcement

	Unit	Labour hours	Net labour (£)	Net material (£)	O'heads/profit (£)	Total (£)
raking	m2	0.15	1.20	0.00	0.18	1.38
curved	m2	0.20	1.60	0.00	0.24	1.84

DESIGNED JOINTS/IN SITU CONCRETE

	Unit	Labour hours	Net labour (£)	Net material (£)	O'heads /profit (£)	Total (£)
E40 DESIGNED JOINTS IN IN SITU CONCRETE						
Labour rate 8.00 per hour						
Expansion joint, Flexcell impregnated fibre based joint filler, formed joint						
12.5mm thick						
not exceeding 150mm wide	m	0.12	0.96	1.94	0.43	3.33
150 to 300mm wide	m	0.18	1.44	2.82	0.64	4.90
300 to 450mm wide	m	0.20	1.60	7.41	1.35	10.36
20mm thick						
not exceeding 150mm wide	m	0.15	1.20	2.66	0.58	4.44
150 to 300mm wide	m	0.20	1.60	3.86	0.82	6.28
300 to 450mm wide	m	0.25	2.00	9.64	1.75	13.39

	Unit	Labour hours	Net labour (£)	Net material (£)	O'heads /profit (£)	Total (£)
E41 WORKED FINISHES ON IN SITU CONCRETE						
Labour rate 8.00 per hour						
Prepare level surfaces of unset concrete						
tamping by mechanical means	m2	0.06	0.48	0.00	0.07	0.55
power floating	m2	0.15	1.20	0.00	0.18	1.38
trowelling	m2	0.23	1.84	0.00	0.28	2.12
Mortice in concrete for rag bolt, grout in cement mortar (1:1), depth						
50mm	nr	0.25	2.00	0.66	0.40	3.06
100mm	nr	0.30	2.40	1.20	0.54	4.14
150mm	nr	0.33	2.64	1.70	0.65	4.99
Mortice in concrete for holding down bolt and plates, grout in cement mortar (1:3), depth						
200mm	nr	0.60	4.80	2.30	1.06	8.16
400mm	nr	0.84	6.72	3.45	1.53	11.70

F

MASONRY

MATERIAL COSTS

	Unit	Material supply (£)	Material waste (%)	Total (£)
MATERIAL COSTS				
Gauged mortar (1:1:6)	m3	62. 43	5. 0	65. 55
Cement mortar (1:3)	m3	70. 38	5. 0	73. 90
Precast concrete dense aggregate blocks (to BS6073)				
Precast concrete blocks				
75mm	m2	5. 33	5. 0	5. 60
100mm	m2	5. 60	5. 0	5. 88
140mm	m2	8. 62	5. 0	9. 05
Hollow				
140mm	m2	8. 73	5. 0	9. 17
190mm	m2	10. 66	5. 0	11. 19
215mm	m2	10. 88	5. 0	11. 42
Thermalite Shield blocks				
75mm	m2	6. 28	5. 0	6. 59
100mm	m2	8. 56	5. 0	8. 99
150mm	m2	12. 83	5. 0	13. 47
215mm	m2	18. 45	5. 0	19. 37
Damp-proof courses				
hessian based 3.8kg/m2	m2	5. 62	5. 0	5. 90
asbestos based 3.8kg	m2	3. 41	5. 0	3. 58
asbestos based with lead core 4.9kg	m2	6. 75	5. 0	7. 09
pitch polymer	m2	6. 75	5. 0	7. 09
Expanded polystyrene insulation				
25mm	m2	3. 07	5. 0	3. 22
50mm	m2	5. 16	5. 0	5. 42

RATES FOR MEASURED WORK

	Unit	Material supply (£)	Material waste (%)	Total (£)
Exmet brick reinforcement, width				
65mm	m	0. 26	5. 0	0. 27
115mm	m	0. 46	5. 0	0. 48
175mm	m	0. 70	5. 0	0. 74
225mm	m	0. 93	5. 0	0. 98
Terracotta air bricks, square hole				
215 x 65mm	nr	1. 85	5. 0	1. 94
215 x 140mm	nr	2. 81	5. 0	2. 95
215 x 215mm	nr	4. 92	5. 0	5. 17
Terracotta air bricks, louvres				
215 x 65mm	nr	1. 18	5. 0	1. 24
215 x 140mm	nr	2. 28	5. 0	2. 39
215 x 215mm	nr	3. 38	5. 0	3. 55
Galvanized cast iron air bricks louvre pattern				
225 x 75mm	nr	3. 15	5. 0	3. 31
225 x 150mm	nr	5. 82	5. 0	6. 11
225 x 225mm	nr	7. 68	5. 0	8. 06
Gas flue block system, type HP range				
recess block (HP1)	nr	3. 39	5. 0	3. 56
cover block (HP2)	nr	4. 80	5. 0	5. 04
straight block, 220mm (HP3)	nr	3. 15	5. 0	3. 31
straight block, 112mm (HP3)	nr	3. 15	5. 0	3. 31
straight block, 72mm (HP3)	nr	3. 15	5. 0	3. 31
vent block (HP3 to 4)	nr	6. 38	5. 0	6. 70
offset block (HP5)	nr	6. 38	5. 0	6. 70
reverse rebate block (HP9)	nr	3. 06	5. 0	3. 21
straight block, 222mm (HP4)	nr	3. 06	5. 0	3. 21
straight block, 112mm (HP4)	nr	3. 06	5. 0	3. 21
straight block, 72mm (HP4)	nr	3. 06	5. 0	3. 21
side offset (HP5)	nr	3. 78	5. 0	3. 97

MATERIAL COSTS

	Unit	Material supply (£)	Material waste (%)	Total (£)
back offset, (HP6)	nr	9.32	5.0	9.79
vertical exit, (HP7)	nr	6.26	5.0	6.57
angled entry/exit, (HP8)	nr	6.26	5.0	6.57
reverse rebate, (HP9)	nr	4.69	5.0	4.92
corbel block (HP10)	nr	5.91	5.0	6.21
lintel block (HP11)	nr	5.62	5.0	5.90

True Flue lining system 200 x
200mm square lining blocks

	Unit	Material supply (£)	Material waste (%)	Total (£)
250mm lining block	m	2.99	5.0	3.14
lintel gather (UL4)	nr	19.48	5.0	20.45
lintel gather (UL5 + U1)	nr	19.88	5.0	20.87
swivel unit (2U or 3U)	nr	8.52	5.0	8.95
bend (4U)	nr	8.52	5.0	8.95
offset (5U) 90mm	nr	4.47	5.0	4.69
offset (6U) 70mm	nr	4.04	5.0	4.24
offset (5U) 183mm	nr	8.70	5.0	9.14
offset (6U) 140mm	nr	7.82	5.0	8.21
milner fireback	nr	26.08	5.0	27.38
natural Portland pot	nr	5.86	5.0	6.15
Terracotta coloured pot	nr	13.03	5.0	13.68

Terracotta chimney pot tapered or
cannon head

	Unit	Material supply (£)	Material waste (%)	Total (£)
300mm	nr	17.85	5.0	18.74
375mm	nr	19.12	5.0	20.08
450mm	nr	21.48	5.0	22.55
600mm	nr	30.86	5.0	32.40
750mm	nr	41.14	5.0	43.20
Rough rubble walling	t	67.56	5.0	70.94
Irregular squared rubble walling	t	84.32	5.0	88.54
Coursed rubble walling	t	105.39	5.0	110.66

Flat section steel arch bars

	Unit	Material supply (£)	Material waste (%)	Total (£)
25 x 10mm	m	1.16	5.0	1.22
50 x 10mm	m	2.26	5.0	2.37
65 x 10mm	m	2.80	5.0	2.94
100 x 10mm	m	4.04	5.0	4.24

RATES FOR MEASURED WORK

	Unit	Material supply (£)	Material waste (%)	Total (£)
Angle section steel arch bars				
40 x 40 x 6mm	m	1.91	5.0	2.01
50 x 50 x 6mm	m	2.42	5.0	2.54
Standard galvanized steel lintels CN102 for 100mm wall				
900mm long	nr	3.00	5.0	3.15
1050mm long	nr	3.59	5.0	3.77
1200mm long	nr	3.94	5.0	4.14
Standard galvanized steel lintels CN7 143mm high for cavity wall				
900mm long	nr	17.94	5.0	18.84
1500mm long	nr	30.96	5.0	32.51
2100mm long	nr	42.52	5.0	44.65
Standard galvanized steel lintels CH8 219mm high for cavity wall				
2700mm long	nr	68.70	5.0	72.14
4200mm long	nr	146.43	5.0	153.75

BRICK/BLOCK WALLING

	Unit	Labour hours	Net labour (£)	Net material (£)	O'heads /profit (£)	Total (£)
F10 BRICK/BLOCK WALLING						
Labour rate 11.50 per hour						
BRICKWORK						
Common bricks BP 130.00 pounds per 1000 in gauged mortar (1:1:6)						
Walls, facework one side						
half brick thick	m2	1.69	19.43	9.30	4.31	33.04
one brick thick	m2	2.85	32.77	19.02	7.77	59.56
one and a half brick thick	m2	3.30	37.95	28.95	10.04	76.94
two brick thick	m2	4.15	47.73	38.67	12.96	99.36
Walls, facework both sides						
half brick thick	m2	1.84	21.16	9.30	4.57	35.03
one brick thick	m2	3.00	34.50	19.02	8.03	61.55
one and a half brick thick	m2	3.54	40.71	28.95	10.45	80.11
two brick thick	m2	4.30	49.45	38.67	13.22	101.34
Skins of hollow walls						
half brick thick	m2	1.54	17.71	9.30	4.05	31.06
one brick thick	m2	2.70	31.05	19.02	7.51	57.58
Honeycomb walls						
half brick thick	m2	1.02	11.73	7.44	2.88	22.05
Dwarf solid walls						
half brick thick	m2	1.21	13.91	9.30	3.48	26.69
Isolated casings						
one brick thick	m2	3.38	38.87	19.02	8.68	66.57
one and a half brick thick	m2	4.05	46.57	28.95	11.33	86.85
two brick thick	m2	4.98	57.27	38.67	14.39	110.33

Common brickwork (cont'd)	Unit	Labour hours	Net labour (£)	Net material (£)	O'heads /profit (£)	Total (£)
Chimney stacks						
one brick thick	m2	3.71	42.66	19.02	9.25	70.93
one and a half brick thick	m2	4.47	51.40	28.95	12.05	92.40
two brick thick	m2	5.48	63.02	38.67	15.25	116.94
Backing to masonry, cutting and building						
one brick thick	m2	4.60	52.90	19.02	10.79	82.71
one and a half brick thick	m2	4.93	56.69	28.95	12.85	98.49
Projections of chimney breasts						
half brick thick	m2	1.92	22.08	9.30	4.71	36.09
one brick thick	m2	3.08	35.42	19.02	8.17	62.61
one and a half brick thick	m2	3.67	42.20	28.95	10.67	81.82
two brick thick	m2	4.50	51.75	38.67	13.56	103.98
Projections of attached piers, plinths, bands and the like						
225 x 112mm	m	0.55	6.33	3.21	1.43	10.97
225 x 225mm	m	1.47	16.91	6.55	3.52	26.98
337 x 225mm	m	2.10	24.15	9.96	5.12	39.23
Bonding ends to existing						
half brick thick	m	0.35	4.02	0.00	0.60	4.62
one brick thick	m	0.50	5.75	0.00	0.86	6.61
one and a half brick thick	m	0.75	8.62	0.00	1.29	9.91
two brick thick	m	1.10	12.65	0.00	1.90	14.55

Common bricks BP 140.00 pounds per 1000 in cement mortar (1:3)

Walls, facework one side						
half brick thick	m2	1.69	19.43	10.07	4.42	33.92
one brick thick	m2	2.85	32.77	20.67	8.02	61.46
one and a half brick thick	m2	3.39	38.98	31.46	10.57	81.01
two brick thick	m2	4.15	47.73	42.00	13.46	103.19

BRICK/BLOCK WALLING

	Unit	Labour hours	Net labour (£)	Net material (£)	O'heads /profit (£)	Total (£)
Walls, facework both sides						
half brick thick	m2	1.84	21.16	10.09	4.69	35.94
one brick thick	m2	3.00	34.50	20.67	8.28	63.45
one and a half brick thick	m2	3.54	40.71	31.46	10.83	83.00
two brick thick	m2	4.30	49.45	42.04	13.72	105.21
Skins of hollow walls						
half brick thick	m2	1.54	17.71	10.09	4.17	31.97
one brick thick	m2	2.70	31.05	20.67	7.76	59.48
Honeycomb walls						
half brick thick	m2	1.02	11.73	8.07	2.97	22.77
Dwarf solid walls						
half brick thick	m2	1.21	13.91	10.09	3.60	27.60
Isolated casings						
one brick thick	m2	3.38	38.87	20.67	8.93	68.47
one and a half brick thick	m2	4.05	46.57	31.46	11.70	89.73
two brick thick	m2	4.98	57.27	41.03	14.74	113.04
Chimney stacks						
one brick thick	m2	3.71	42.66	20.67	9.50	72.83
one and a half brick thick	m2	4.47	51.40	31.46	12.43	95.29
two brick thick	m2	5.48	63.02	41.03	15.61	119.66
Backing to masonry, cutting and building						
one brick thick	m2	4.10	47.15	20.67	10.17	77.99
one and a half brick thick	m2	4.93	56.69	31.46	13.22	101.37
Projections of chimney breasts						
half brick thick	m	1.92	22.08	10.09	4.83	37.00
one brick thick	m	3.08	35.42	20.67	8.41	64.50
one and a half brick thick	m	3.67	42.20	31.06	10.99	84.25
two brick thick	m	4.50	51.75	42.04	14.07	107.86

111

Common brickwork (cont'd)	Unit	Labour hours	Net labour (£)	Net material (£)	O'heads /profit (£)	Total (£)
Projections of attached piers, plinths, bands and the like						
225 x 112mm	m	0.55	6.33	3.50	1.47	11.30
225 x 225mm	m	1.47	16.91	7.15	3.61	27.67
337 x 225mm	m	2.10	24.15	10.88	5.25	40.28
Bonding ends to existing						
half brick thick	m	0.35	4.02	0.00	0.60	4.62
one brick thick	m	0.50	5.75	0.00	0.86	6.61
one and a half brick thick	m	0.75	8.62	0.00	1.29	9.91
two brick thick	m	1.10	12.65	0.00	1.90	14.55
Common bricks BP 150.00 pounds per 1000 in cement mortar (1:3)						
Walls, facework one side						
half brick thick	m2	1.69	19.43	10.69	4.52	34.64
one brick thick	m2	2.85	32.77	21.85	8.19	62.81
one and a half brick thick	m2	3.39	38.98	33.24	10.83	83.05
two brick thick	m2	4.15	47.73	44.60	13.85	106.18
Walls, facework both sides						
half brick thick	m2	1.84	21.16	10.69	4.78	36.63
one brick thick	m2	3.00	34.50	21.85	8.45	64.80
one and a half brick thick	m2	3.54	40.71	33.24	11.09	85.04
two brick thick	m2	4.30	49.45	44.60	14.11	108.16
Skins of hollow walls						
half brick thick	m2	1.54	17.71	10.69	4.26	32.66
one brick thick	m2	2.70	31.05	21.85	7.93	60.83
Honeycomb walls						
half brick thick	m2	1.02	11.73	8.53	3.04	23.30
Dwarf solid walls						
half brick thick	m2	1.21	13.91	10.69	3.69	28.29

BRICK/BLOCK WALLING

	Unit	Labour hours	Net labour (£)	Net material (£)	O'heads /profit (£)	Total (£)
Isolated casings						
one brick thick	m2	3.38	38.87	21.85	9.11	69.83
one and a half brick thick	m2	4.05	46.57	33.24	11.97	91.78
two brick thick	m2	4.98	57.27	44.40	15.25	116.92
Chimney stacks						
one brick thick	m2	3.71	42.66	21.85	9.68	74.19
one and a half brick thick	m2	4.47	51.40	33.24	12.70	97.34
two brick thick	m2	5.48	63.02	44.40	16.11	123.53
Backing to masonry, cutting and building						
one brick thick	m2	4.10	47.15	21.85	10.35	79.35
one and a half brick thick	m2	4.93	56.69	33.24	13.49	103.42
Projections of chimney breasts						
half brick thick	m2	1.92	22.08	10.69	4.92	37.69
one brick thick	m2	3.08	35.42	21.85	8.59	65.86
one and a half brick thick	m2	3.67	42.20	33.24	11.32	86.76
two brick thick	m2	4.50	51.75	44.40	14.42	110.57
Projections of attached piers, plinths, bands and the like						
225 x 112mm	m	0.55	6.33	3.71	1.51	11.55
225 x 225mm	m	1.47	16.91	7.56	3.67	28.14
337 x 225mm	m	2.10	24.15	11.53	5.35	41.03
Bonding ends to existing						
half brick thick	m	0.35	4.02	0.00	0.60	4.62
one brick thick	m	0.50	5.75	0.00	0.86	6.61
one and a half brick thick	m	0.75	8.62	0.00	1.29	9.91
two brick thick	m	1.10	12.65	0.00	1.90	14.55

RATES FOR MEASURED WORK

	Unit	Labour hours	Net labour (£)	Net material (£)	O'heads /profit (£)	Total (£)
Class B engineering bricks BP 240.00 pounds per 1000 in cement mortar (1:3)						
Walls, facework one side						
half brick thick	m2	1.85	21.28	16.09	5.61	42.98
one brick thick	m2	3.07	35.30	32.47	10.17	77.94
one and a half brick thick	m2	3.65	41.98	49.28	13.69	104.95
two brick thick	m2	4.53	52.10	64.63	17.51	134.24
Walls, facework both sides						
half brick thick	m2	2.00	23.00	16.09	5.86	44.95
one brick thick	m2	3.22	37.03	32.47	10.42	79.92
one and a half brick thick	m2	3.80	43.70	49.28	13.95	106.93
two brick thick	m2	4.68	53.82	64.63	17.77	136.22
Skins of hollow walls						
half brick thick	m2	1.70	19.55	16.09	5.35	40.99
one brick thick	m2	2.92	33.58	32.47	9.91	75.96
Isolated coverings						
one brick thick	m2	3.62	41.63	32.47	11.12	85.22
one and a half brick thick	m2	4.36	50.14	49.26	14.91	114.31
two brick thick	m2	5.38	61.87	64.63	18.97	145.47
Chimney stacks						
one brick thick	m2	4.00	46.00	32.47	11.77	90.24
one and a half brick thick	m2	4.75	54.62	49.26	15.58	119.46
two brick thick	m2	5.88	67.62	64.63	19.84	152.09
Backing to masonry, cutting and building						
one brick thick	m	4.45	51.18	32.47	12.55	96.20
one and a half brick thick	m	5.45	62.68	49.26	16.79	128.73

114

BRICK/BLOCK WALLING

	Unit	Labour hours	Net labour (£)	Net material (£)	O'heads /profit (£)	Total (£)
Projections of chimney breasts						
half brick thick	m	2.10	24.15	16.09	6.04	46.28
one brick thick	m	3.34	38.41	32.47	10.63	81.51
one and a half brick thick	m	4.18	48.07	49.26	14.60	111.93
two brick thick	m	4.96	57.04	64.63	18.25	139.92
Projections of attached piers, plinths, bands and the like						
225 x 112mm	m	0.71	8.16	5.60	2.06	15.82
225 x 225mm	m	1.70	19.55	11.28	4.62	35.45
337 x 225mm	m	2.32	26.68	17.10	6.57	50.35
Bonding ends to existing						
half brick thick	m	0.35	4.02	0.00	0.60	4.62
one brick thick	m	0.50	5.75	0.00	0.86	6.61
one and a half brick thick	m	0.75	8.62	0.00	1.29	9.91
two brick thick	m	1.10	12.65	0.00	1.90	14.55
Engineering bricks BP 350.00 pounds per 1000 in gauged mortar (1:1:6)						
Walls, facework one side						
half brick thick	m2	1.85	21.28	22.50	6.57	50.35
half brick skin of hollow wall	m2	3.07	35.30	22.37	8.65	66.32
one brick thick	m2	3.65	41.98	44.98	13.04	100.00
Walls, facework both sides						
half brick thick	m2	2.00	23.00	22.50	6.83	52.33
one brick thick	m2	3.65	41.98	44.98	13.04	100.00
Bonding ends to existing						
half brick thick	m	0.35	4.02	0.00	0.60	4.62
one brick thick	m	0.50	5.75	0.00	0.86	6.61
one and a half brick thick	m	0.75	8.62	0.00	1.29	9.91
two brick thick	m	1.10	12.65	0.00	1.90	14.55

RATES FOR MEASURED WORK

	Unit	Labour hours	Net labour (£)	Net material (£)	O'heads /profit (£)	Total (£)
Facings BP 400.00 pounds per 1000 in gauged mortar (1:1:6)						
Walls, facework one side						
half brick thick	m2	2.04	23.46	27.19	7.60	58.25
half brick skin of hollow wall	m2	2.14	24.61	27.19	7.77	59.57
one brick thick	m2	3.20	36.80	50.88	13.15	100.83
Walls, facework both sides						
half brick thick	m2	2.31	26.57	27.19	8.06	61.82
one brick thick	m2	3.47	39.91	50.88	13.62	104.41
Bonding ends to existing						
half brick thick	m	0.35	4.02	0.00	0.60	4.62
one brick thick	m	0.50	5.75	0.00	0.86	6.61
one and a half brick thick	m	0.75	8.62	0.00	1.29	9.91
two brick thick	m	1.10	12.65	0.00	1.90	14.55
Facings BP 500.00 pounds per 1000 in gauged mortar (1:1:6)						
Walls, facework one side						
half brick thick	m2	2.04	23.46	33.19	8.50	65.15
half brick skin of hollow wall	m2	2.14	24.61	33.19	8.67	66.47
one brick thick	m2	3.20	36.80	62.68	14.92	114.40
Walls, facework both sides						
half brick thick	m2	2.31	26.57	33.19	8.96	68.72
one brick thick	m2	3.47	39.91	62.68	15.39	117.98
Bonding ends to existing						
half brick thick	m	0.35	4.02	0.00	0.60	4.62
one brick thick	m	0.50	5.75	0.00	0.86	6.61
one and a half brick thick	m	0.75	8.62	0.00	1.29	9.91
two brick thick	m	1.10	12.65	0.00	1.90	14.55

BRICK/BLOCK WALLING

	Unit	Labour hours	Net labour (£)	Net material (£)	O'heads /profit (£)	Total (£)
BLOCKWORK						
Precast concrete dense aggregate block to BS6073 in gauged mortar (1:1:6)						
Solid blocks in walls and partitions, thickness						
75mm	m2	0.98	11.27	5.90	2.58	19.75
100mm	m2	1.14	13.11	6.13	2.89	22.13
140mm	m2	1.32	15.18	9.59	3.72	28.49
Hollow blocks in walls and partitions, thickness						
140mm	m2	1.43	16.45	9.72	3.93	30.10
190mm	m2	1.76	20.24	12.00	4.84	37.08
215mm	m2	2.02	23.23	12.29	5.33	40.85
Solid blocks in skins of hollow walls, thickness						
75mm	m2	1.12	12.88	5.90	2.82	21.60
100mm	m2	1.24	14.26	6.31	3.09	23.66
140mm	m2	1.47	16.91	9.59	3.97	30.47
Hollow blocks in skins of hollow walls, thickness						
140mm	m2	1.56	17.94	9.72	4.15	31.81
190mm	m2	1.90	21.85	12.00	5.08	38.93
215mm	m2	2.15	24.72	12.29	5.55	42.56
Solid blocks in piers and chimney stacks, thickness						
75mm	m2	1.28	14.72	5.90	3.09	23.71
100mm	m2	1.40	16.10	6.13	3.33	25.56
140mm	m2	1.68	19.32	9.59	4.34	33.25

Blockwork (cont'd)	Unit	Labour hours	Net labour (£)	Net material (£)	O'heads /profit (£)	Total (£)
Hollow blocks in piers and chimney stacks, thickness						
140mm	m2	1.68	19.32	9.72	4.36	33.40
190mm	m2	2.04	23.46	12.00	5.32	40.78
215mm	m2	2.37	27.26	12.29	5.93	45.48
Solid blocks in isolated casings, thickness						
75mm	m2	1.42	16.33	5.90	3.33	25.56
100mm	m2	1.90	21.85	6.31	4.22	32.38
140mm	m2	2.15	24.72	9.59	5.15	39.46
Hollow blocks in isolated casings, thickness						
140mm	m2	1.82	20.93	9.72	4.60	35.25
190mm	m2	2.30	26.45	12.00	5.77	44.22
215mm	m2	2.48	28.52	12.29	6.12	46.93
Extra for fair face and flush pointing						
one side	m2	0.16	1.84	0.00	0.28	2.12
both sides	m2	0.33	3.80	0.00	0.57	4.37
Bonding ends of blockwork to brickwork in alternate course, width						
75mm	m	0.30	3.45	0.00	0.52	3.97
100mm	m	0.38	4.37	0.00	0.66	5.03
140mm	m	0.44	5.06	0.00	0.76	5.82
190mm	m	0.55	6.33	0.00	0.95	7.28
215mm	m	0.62	7.13	0.00	1.07	8.20

BRICK/BLOCK WALLING

	Unit	Labour hours	Net labour (£)	Net material (£)	O'heads /profit (£)	Total (£)
Thermalite Shield blocks in cement mortar (1:3)						
In walls and partitions, thickness						
75mm	m2	0.87	10.01	7.20	2.58	19.79
100mm	m2	1.03	11.85	9.84	3.25	24.94
150mm	m2	1.18	13.57	14.67	4.24	32.48
215mm	m2	1.32	15.18	21.17	5.45	41.80
In skins of hollow walls, thickness						
75mm	m2	1.00	11.50	7.20	2.80	21.50
100mm	m2	1.10	12.65	9.84	3.37	25.86
150mm	m2	1.31	15.07	14.67	4.46	34.20
215mm	m2	1.98	22.77	21.17	6.59	50.53
In piers and chimney stacks, thickness						
75mm	m2	1.18	13.57	7.20	3.12	23.89
100mm	m2	1.28	14.72	9.84	3.68	28.24
150mm	m2	1.52	17.48	14.67	4.82	36.97
215mm	m2	1.97	22.66	21.17	6.57	50.40
In isolated casings, thickness						
75mm	m2	1.92	22.08	7.20	4.39	33.67
100mm	m2	1.74	20.01	9.84	4.48	34.33
150mm	m2	1.97	22.66	14.67	5.60	42.93
215mm	m2	2.02	23.23	21.17	6.66	51.06
Extra for fair face and flush pointing						
one side	m2	0.14	1.61	0.00	0.24	1.85
both sides	m2	0.27	3.11	0.00	0.47	3.58

119

Blockwork (cont'd)	Unit	Labour hours	Net labour (£)	Net material (£)	O'heads /profit (£)	Total (£)
Bonding ends of blockwork to brickwork in alternate course, width						
75mm	m	0.28	3.22	0.00	0.48	3.70
100mm	m	0.31	3.56	0.00	0.53	4.09
150mm	m	0.38	4.37	0.00	0.66	5.03
215mm	m	0.54	6.21	0.00	0.93	7.14
Thermalite Shield blocks in gauged mortar (1:1:6)						
In walls and partitions, thickness						
75mm	m2	0.87	10.01	7.22	2.58	19.81
100mm	m2	1.03	11.85	9.84	3.25	24.94
150mm	m2	1.18	13.57	14.67	4.24	32.48
215mm	m2	1.32	15.18	21.17	5.45	41.80
In skins of hollow walls, thickness						
75mm	m2	1.00	11.50	7.22	2.81	21.53
100mm	m2	1.10	12.65	9.84	3.37	25.86
150mm	m2	1.31	15.07	14.67	4.46	34.20
215mm	m2	1.98	22.77	21.17	6.59	50.53
In piers and chimney stacks, thickness						
75mm	m2	1.18	13.57	7.22	3.12	23.91
100mm	m2	1.28	14.72	9.84	3.68	28.24
150mm	m2	1.52	17.48	14.67	4.82	36.97
215mm	m2	1.97	22.66	21.17	6.57	50.40
In isolated casings, thickness						
75mm	m2	1.92	22.08	7.22	4.39	33.69
100mm	m2	1.74	20.01	9.84	4.48	34.33
150mm	m2	1.97	22.66	14.67	5.60	42.93
215mm	m2	2.02	23.23	21.17	6.66	51.06

BRICK/BLOCK WALLING

	Unit	Labour hours	Net labour (£)	Net material (£)	O'heads /profit (£)	Total (£)
Extra for fair face and flush pointing						
one side	m2	0.14	1.61	0.00	0.24	1.85
both sides	m2	0.27	3.11	0.00	0.47	3.58
Bonding ends of blockwork to brickwork in alternate course, width						
75mm	m	0.28	3.22	0.00	0.48	3.70
100mm	m	0.33	3.80	0.00	0.57	4.37
150mm	m	0.38	4.37	0.00	0.66	5.03
215mm	m	0.54	6.21	0.00	0.93	7.14

	Unit	Labour hours	Net labour (£)	Net material (£)	O'heads /profit (£)	Total (£)
F20 NATURAL STONE RUBBLE WALLING						
Labour rate 8.00 per hour						
Random rubble walling, laid dry, thickness						
300mm	m2	3.00	24.00	41.29	9.79	75.08
450mm	m2	3.25	26.00	61.93	13.19	101.12
Random rubble walling, laid dry, battered both faces, average thickness 450mm	m2	3.75	30.00	61.93	13.79	105.72
Random rubble walling in gauged mortar (1:1:6), thickness						
300mm	m2	1.50	12.00	44.35	8.45	64.80
450mm	m2	2.25	18.00	68.24	12.94	99.18
550mm	m2	2.75	22.00	80.48	15.37	117.85
Random rubble walling in cement mortar (1:3)						
300mm	m2	1.50	12.00	42.91	8.24	63.15
450mm	m2	2.25	18.00	66.11	12.62	96.73
550mm	m2	2.75	22.00	80.81	15.42	118.23
Irregular coursed rubble walling in gauged mortar (1:1:6)						
300mm	m2	1.95	15.60	58.61	11.13	85.34
450mm	m2	2.92	23.36	84.03	16.11	123.50
550mm	m2	3.58	28.64	99.77	19.26	147.67
Irregular coursed rubble walling in cement mortar (1:3)						
300mm	m2	1.95	15.60	57.17	10.92	83.69
450mm	m2	2.92	23.36	81.90	15.79	121.05
550mm	m2	3.58	28.64	100.10	19.31	148.05

NATURAL STONE RUBBLE WALLING

	Unit	Labour hours	Net labour (£)	Net material (£)	O'heads /profit (£)	Total (£)
Coursed rubble walling in gauged mortar (1:1:6)						
300mm	m2	1.60	12.80	72.61	12.81	98.22
450mm	m2	2.40	19.20	103.90	18.46	141.57
550mm	m2	2.93	23.44	124.07	22.13	169.64
Coursed rubble walling in cement mortar (1:3)						
300mm	m2	1.60	12.80	71.17	12.60	96.57
450mm	m2	2.40	19.20	101.77	18.15	139.12
550mm	m2	2.93	23.44	124.40	22.18	170.02

	Unit	Labour hours	Net labour (£)	Net material (£)	O'heads /profit (£)	Total (£)

F30 ACCESSORIES/SUNDRY ITEMS FOR MASONRY

Labour rate 8.00 per hour

Form 50mm cavity between skin of
hollow walls, 3 ties per m2

galvanized steel butterfly ties	m2	0.10	0.80	0.40	0.18	1.38
galvanized steel twisted ties	m2	0.10	0.80	0.66	0.22	1.68
stainless steel twisted ties	m2	0.10	0.80	0.58	0.21	1.59

Form 75mm cavity between skin of
hollow walls, 3 ties per m2

galvanized steel butterfly ties	m2	0.10	0.80	0.40	0.03	1.23
galvanized steel twisted ties	m2	0.10	0.80	0.66	0.22	1.68
stainless steel ties	m2	0.10	0.80	0.58	0.21	1.59

Hessian based bitumen damp-proof course in gauged mortar (1:1:6)

Horizontal width

over 225mm	m2	0.35	2.80	6.04	1.33	10.17
not exceeding 225mm	m2	0.60	4.80	6.04	1.63	12.47

Vertical width

over 225mm	m2	0.40	3.20	6.04	1.39	10.63
not exceeding 225mm	m2	0.20	1.60	6.04	1.15	8.79

Asbestos based bitumen damp-proof course in gauged mortar (1:1:6)

Horizontal width

over 225mm	m2	0.35	2.80	3.72	0.98	7.50
not exceeding 225mm	m2	0.60	4.80	3.72	1.28	9.80

ACCESSORIES FOR WALLING

	Unit	Labour hours	Net labour (£)	Net material (£)	O'heads /profit (£)	Total (£)
Vertical width						
over 225mm	m2	0.40	3.20	3.72	1.04	7.96
not exceeding 225mm	m2	0.70	5.60	3.72	1.40	10.72
Pitch polymer damp-proof course in gauged mortar (1:1:6)						
Horizontal width						
over 225mm	m2	0.35	2.80	7.23	1.50	11.53
not exceeding 225mm	m2	0.60	4.80	7.23	1.80	13.83
Vertical width						
over 225mm	m2	0.40	3.20	7.23	1.56	11.99
not exceeding 225mm	m2	0.70	5.60	7.23	1.92	14.75
Asbestos based with lead core damp-proof course in gauged mortar (1:1:6)						
Horizontal width						
over 225mm	m2	0.40	3.20	7.23	1.56	11.99
not exceeding 225mm	m2	0.70	5.60	7.23	1.92	14.75
Vertical width						
over 225mm	m2	0.45	3.60	7.23	1.62	12.45
not exceeding 225mm	m2	0.75	6.00	7.23	1.98	15.21
Three coats Synthaprufe waterproofing liquid brushed on surfaces						
vertically	m2	0.42	3.36	2.99	0.95	7.30
horizontally	m2	0.33	2.64	2.99	0.84	6.47
Cavity wall insulation expanded polystyrene sheets						
25mm	m2	0.16	1.28	3.22	0.68	5.18
50mm	m2	0.16	1.28	5.42	1.00	7.70

RATES FOR MEASURED WORK

	Unit	Labour hours	Net labour (£)	Net material (£)	O'heads /profit (£)	Total (£)
Galvanized brick reinforcement, 'Exmet' width						
65mm	m	0.07	0.56	0.27	0.12	0.95
115mm	m	0.09	0.72	0.49	0.18	1.39
175mm	m	0.12	0.96	0.73	0.25	1.94
225mm	m	0.15	1.20	0.97	0.33	2.50
Point frames with mastic						
one side	m	0.12	0.96	0.23	0.18	1.37
two sides	m	0.20	1.60	0.25	0.28	2.13
Point frames with polysulphide sealant						
one side	m	0.12	0.96	1.21	0.33	2.50
two sides	m	0.20	1.60	1.90	0.53	4.03
Slate damp-proof course bedded and pointed in cement mortar (1:3) over 225mm wide						
single course	m2	0.40	3.20	23.86	4.06	31.12
double course	m2	0.20	1.60	47.36	7.34	56.30
Slate damp-proof course bedded and pointed in cement mortar (1:3) not exceeding 225mm wide						
single course	m2	0.45	3.60	23.86	4.12	31.58
double course	m2	0.80	6.40	47.36	8.06	61.82
Rake out joints of brickwork for flashing and point up on completion						
horizontal	m	0.09	0.72	0.14	0.13	0.99
stepped	m	0.17	1.36	0.18	0.23	1.77

ACCESSORIES FOR WALLING

	Unit	Labour hours	Net labour (£)	Net material (£)	O'heads /profit (£)	Total (£)
True Flue lining system						
Typex High HP range, bedded in refractory mortar (1:2:5), building into brickwork or block walls						
recess block (HP1)	nr	0.18	1.44	3.55	0.75	5.74
cover block (HP2)	nr	0.18	1.44	5.04	0.97	7.45
222mm standard block (HP3)	nr	0.18	1.44	3.30	0.71	5.45
112mm standard block (HP3)	nr	0.18	1.44	3.30	0.71	5.45
72mm standard block (HP3)	nr	0.18	1.44	3.30	0.71	5.45
vent block (HP3-4)	nr	0.18	1.44	7.00	1.27	9.71
222mm standard block (HP4)	nr	0.18	1.44	3.21	0.70	5.35
112mm standard block (HP4)	nr	0.18	1.44	3.21	0.70	5.35
72mm standard block (HP4)	nr	0.18	1.44	3.21	0.70	5.35
side offset (HP5)	nr	0.20	1.60	3.96	0.83	6.39
back offset (HP6)	nr	0.20	1.60	9.78	1.71	13.09
vertical exit (HP7)	nr	0.25	2.00	6.57	1.29	9.86
angled entry/exit (HP8)	nr	0.25	2.00	6.57	1.29	9.86
reverse rebate (HP9)	nr	0.25	2.00	4.92	1.04	7.96
corbel block (HP10)	nr	0.30	2.40	6.20	1.29	9.89
lintel block (HP11)	nr	0.30	2.40	5.90	1.24	9.54
True Flue 200 x 200mm square lining, bedded in refractory mortar (1:2:5), building into brickwork or blockwork walls						
lining block 250mm high	m	0.18	1.44	3.14	0.69	5.27
lintel gather (UL4)	nr	0.30	2.40	20.87	3.49	26.76
lintel gather with lintel (UL5 + U1)	nr	0.40	3.20	20.87	3.61	27.68
swivel unit (2U or 3U)	nr	0.25	2.00	8.95	1.64	12.59
bend (4U)	nr	0.18	1.44	8.95	1.56	11.95
offset (5U) 90mm	nr	0.18	1.44	4.69	0.92	7.05
offset (6U) 70mm	nr	0.18	1.44	4.24	0.85	6.53
offset (5U) 183mm	nr	0.18	1.44	9.14	1.59	12.17
offset (6U) 140mm	nr	0.18	1.44	8.21	1.45	11.10
Milner fire back	nr	0.50	4.00	27.38	0.78	32.16
natural Portland coloured pot	nr	1.50	12.00	6.15	2.72	20.87
terracotta coloured pot	nr	1.50	12.00	13.68	3.85	29.53

RATES FOR MEASURED WORK

	Unit	Labour hours	Net labour (£)	Net material (£)	O'heads /profit (£)	Total (£)

Chimney pots

Terracotta chimney pot, tapered
or cannon head setting and
flaunching in cement mortar,
185mm diameter, height

	Unit	Labour hours	Net labour (£)	Net material (£)	O'heads /profit (£)	Total (£)
300mm	nr	1.30	10.40	18.74	4.37	33.51
375mm	nr	1.50	12.00	20.08	4.81	36.89
450mm	nr	1.80	14.40	22.55	5.54	42.49
600mm	nr	2.00	16.00	32.40	7.26	55.66
750mm	nr	2.40	19.20	43.20	9.36	71.76

Air bricks

Form opening in cavity wall for
air brick, seal cavity with slates
in cement mortar (1:3), size

	Unit	Labour hours	Net labour	Net material	O'heads /profit	Total
225 x 75mm	nr	0.30	2.40	2.33	0.71	5.44
225 x 150mm	nr	0.42	3.36	3.01	0.96	7.33
225 x 225mm	nr	0.54	4.32	4.35	1.30	9.97

Terracotta louvre pattern air
bricks, size

	Unit	Labour hours	Net labour	Net material	O'heads /profit	Total
215 x 65mm	nr	0.10	0.80	1.94	0.41	3.15
215 x 140mm	nr	0.10	0.80	2.95	0.56	4.31
215 x 215mm	nr	0.10	0.80	5.16	0.89	6.85

Terracotta square hole pattern
air bricks, size

	Unit	Labour hours	Net labour	Net material	O'heads /profit	Total
215 x 65mm	nr	0.10	0.80	1.23	0.30	2.33
215 x 140mm	nr	0.10	0.80	2.39	0.48	3.67
215 x 215mm	nr	0.10	0.80	3.54	0.65	4.99

Galvanized cast iron louvre
pattern air bricks, size

	Unit	Labour hours	Net labour	Net material	O'heads /profit	Total
225 x 75mm	nr	0.12	0.96	3.30	0.64	4.90
225 x 150mm	nr	0.12	0.96	6.11	1.06	8.13
225 x 225mm	nr	0.12	0.96	8.06	1.35	10.37

ACCESSORIES FOR WALLING

	Unit	Labour hours	Net labour (£)	Net material (£)	O'heads /profit (£)	Total (£)
Bars and lintels						
Flat section steel arch bars						
25 x 10mm	m	0.15	1.20	1.21	0.36	2.77
50 x 10mm	m	0.18	1.44	2.37	0.57	4.38
65 x 10mm	m	0.22	1.76	2.94	0.70	5.40
100 x 10mm	m	0.25	2.00	4.24	0.94	7.18
Angle section steel arch bars						
40 x 40 x 6mm	m	0.25	2.00	2.00	0.60	4.60
50 x 50 x 6mm	m	0.45	3.60	2.54	0.92	7.06
Standard galvanized steel lintels to 100mm thick wall, 143mm high						
900mm long	m	0.15	1.20	3.15	0.65	5.00
1500mm long	m	0.20	1.60	3.77	0.81	6.18
2100mm long	m	0.25	2.00	4.13	0.92	7.05
Standard galvanized steel lintels to 256mm thick walls, 143mm high						
900mm long	m	0.20	1.60	18.74	3.05	23.39
1500mm long	m	0.25	2.00	32.51	5.18	39.69
2100mm long	m	0.30	2.40	44.75	7.07	54.22
Standard galvanized steel lintels to 256mm thick walls, 219mm high						
2700mm long	m	0.75	6.00	72.14	11.72	89.86
4200mm long	m	1.00	8.00	153.75	24.26	186.01

	Unit	Labour hours	Net labour (£)	Net material (£)	O'heads /profit (£)	Total (£)

F31 PRECAST CONCRETE

Labour rate 8.00 per hour

Lintels (21N/mm2), bedded and
pointed in cement mortar (1:3),
size

65 x 100mm, length

1.35m	nr	0.30	2.40	5.16	1.13	8.69
2.25m	nr	0.30	2.40	8.59	1.65	12.64

65 x 150mm, length

1.35m	nr	0.35	2.80	6.63	1.41	10.84
2.25m	nr	0.40	3.20	11.06	2.14	16.40

65 x 220mm, length

1.35m	nr	0.40	3.20	10.81	2.10	16.11
2.25m	nr	0.50	4.00	18.01	3.30	25.31

Copings (30N/mm2) weathered and
throated, bedded in cement mortar
(1:3)

75 x 300 x 900mm	nr	0.62	4.96	6.20	1.67	12.83
75 x 300 x 600mm	nr	0.80	6.40	4.31	1.61	12.32

Copings (30N/mm2) twice weathered
and twice throated, bedded in
cement mortar (1:3)

75 x 300 x 600mm	nr	0.62	4.96	4.31	1.39	10.66
60 x 160 x 450mm	nr	0.80	6.40	2.83	1.38	10.61

G

STRUCTURAL/CARCASSING/METAL/TIMBER

Spon's Budget Estimating Handbook
2nd Edition

Tweeds Chartered Quantity Surveyors, UK

- provides back-of-the envelope calculations
- improved design therefore easier to use
- fine-tuned to suit market's needs

This is the only guide to concentrate entirely on approximate estimating. It contains a broad range of information to help the developer, quantity surveyor, engineer, architect and landscape architect to produce approximate costings to enable early decisions to be taken on the viability of proposed schemes.

This book provides a one-stop reference point in preparing estimates, so saving valuable time and increasing the quality of cost estimates.

Contents: Foreword. Preface to the first edition. Preface. Introduction. **Part One: Building work.** Square metre prices. Elemental costs. Composite rates. **Part Two: Civil engineering work.** Principal rates. Composite rates. Project costs. **Part Three: Mechanical and electrical work.** Square metre prices. Principal rates. **Part Four: Reclamation and landscaping.** Principal rates. Maintenance. **Part Five: Alterations and repairs.** Principal rates. **Part Six: General data.** Life cycle costing. The development process. Professional fees. Construction indices. Rebuilding costs. Index.

April 1994: 234x120: 272pp, 1 line illus
Hardback: 0-419-19250-6: £39.00

For further information and to order please contact: The Promotion Dept.,
E & F N Spon, 2-6 Boundary Row, London SE1 8HN
Tel: 071 865 0066 Fax: 071 522 9623

E & F N Spon
An imprint of Chapman & Hall

MATERIAL COSTS

	Unit	Material supply (£)	Material waste (%)	Total (£)
MATERIAL COSTS				
Sawn softwood untreated				
25 x 50mm	m	0.31	5.0	0.33
25 x 75mm	m	0.48	5.0	0.50
25 x 100mm	m	0.61	5.0	0.64
25 x 150mm	m	0.95	5.0	1.00
38 x 50mm	m	0.50	5.0	0.53
38 x 75mm	m	0.76	5.0	0.80
38 x 100mm	m	1.03	5.0	1.08
50 x 50mm	m	0.56	5.0	0.59
50 x 75mm	m	0.76	5.0	0.80
50 x 100mm	m	1.01	5.0	1.06
50 x 125mm	m	1.25	5.0	1.31
50 x 150mm	m	1.53	5.0	1.61
75 x 75mm	m	1.28	5.0	1.34
75 x 100mm	m	1.71	5.0	1.80
75 x 125mm	m	2.22	5.0	2.33
75 x 150mm	m	2.55	5.0	2.68
Plywood, marine quality				
12mm	m2	9.77	5.0	10.26
18mm	m2	14.31	5.0	15.03
25mm	m2	19.87	5.0	20.86
Rolled steel joists, size				
127 x 76mm	t	533.00	2.5	546.33
127 x 114mm	t	533.00	2.5	546.33
152 x 89mm	t	533.00	2.5	546.33
152 x 127mm	t	533.00	2.5	546.33
Universal columns				
152 x 152mm	t	533.00	2.5	546.33
203 x 203mm	t	533.00	2.5	546.33
254 x 254mm	t	533.00	2.5	546.33

RATES FOR MEASURED WORK

	Unit	Material supply (£)	Material waste (%)	Total (£)
Angles				
200 x 150mm	m	30.91	2.5	31.68
150 x 90mm	m	14.92	2.5	15.29
125 x 75mm	m	11.19	2.5	11.47
100 x 65mm	m	6.93	2.5	7.10
75 x 50mm	m	5.33	2.5	5.46

STRUCTURAL STEEL FRAMING

	Unit	Labour hours	Net labour (£)	Net material (£)	O'heads /profit (£)	Total (£)

G10 STRUCTURAL STEEL FRAMING

Labour rate 8.00 per hour

The involvement of a contractor carrying out small works will probably be limited to fixing rolled steel joists and columns in openings. These rates are typical but attention should be paid to the material cost which can vary widely

Steel to BS4360 Grade 43

Rolled steel joists fixed 2 to 3m above ground level

127 x 76mm	m	0. 90	7. 20	8. 72	2. 39	18. 31
127 x 114mm	m	1. 15	9. 20	15. 86	3. 76	28. 82
152 x 89mm	m	1. 05	8. 40	9. 11	2. 63	20. 14
152 x 127mm	m	1. 25	10. 00	19. 83	4. 47	34. 30

Universal columns

152 x 152mm	m	1. 30	10. 40	19. 72	4. 52	34. 64
203 x 203mm	m	1. 75	14. 00	45. 84	8. 98	68. 82
254 x 254mm	m	2. 50	20. 00	89. 01	16. 35	125. 36

Angles fixing up to 3m high above ground level

200 x 150mm	m	1. 35	10. 80	31. 68	6. 37	48. 85
150 x 90mm	m	0. 95	7. 60	15. 29	3. 43	26. 32
125 x 75mm	m	0. 48	3. 84	11. 47	2. 30	17. 61
100 x 65mm	m	0. 35	2. 80	7. 10	1. 48	11. 38
75 x 50mm	m	0. 28	2. 24	5. 46	1. 16	8. 86

	Unit	Labour hours	Net labour (£)	Net material (£)	O'heads /profit (£)	Total (£)

G20 CARPENTRY/TIMBER

Labour rate 8.00 per hour

CARCASSING

Sawn softwood, untreated

Floors

50 x 100mm	m	0.22	1.76	1.06	0.42	3.24
50 x 125mm	m	0.25	2.00	1.31	0.50	3.81
50 x 150mm	m	0.27	2.16	1.60	0.56	4.32
75 x 125mm	m	0.28	2.24	2.33	0.69	5.26
75 x 150mm	m	0.30	2.40	2.67	0.76	5.83

Partitions

38 x 75mm	m	0.32	2.56	0.80	0.50	3.86
38 x 100mm	m	0.34	2.72	1.08	0.57	4.37
50 x 75mm	m	0.34	2.72	0.80	0.53	4.05
50 x 100mm	m	0.34	2.72	1.06	0.57	4.35

Flat roofs

38 x 100mm	m	0.15	1.20	1.08	0.34	2.62
50 x 75mm	m	0.17	1.36	0.80	0.32	2.48
50 x 100mm	m	0.18	1.44	1.06	0.38	2.88
50 x 125mm	m	0.19	1.52	1.31	0.42	3.25
50 x 150mm	m	0.20	1.60	1.60	0.48	3.68
75 x 100mm	m	0.20	1.60	1.80	0.51	3.91
75 x 125mm	m	0.26	2.08	2.33	0.66	5.07

Pitched roofs

38 x 100mm	m	0.22	1.76	1.08	0.43	3.27
50 x 75mm	m	0.24	1.92	0.80	0.41	3.13
50 x 100mm	m	0.25	2.00	1.06	0.46	3.52
50 x 125mm	m	0.26	2.08	1.31	0.51	3.90
50 x 150mm	m	0.27	2.16	1.60	0.56	4.32
75 x 100mm	m	0.27	2.16	1.80	0.59	4.55
75 x 125mm	m	0.38	3.04	2.33	0.81	6.18
75 x 150mm	m	0.45	3.60	2.67	0.94	7.21

CARPENTRY/TIMBER

	Unit	Labour hours	Net labour (£)	Net material (£)	O'heads /profit (£)	Total (£)
Kerb, bearer						
25 x 75mm	m	0.12	0.96	0.51	0.22	1.69
25 x 100mm	m	0.16	1.28	0.64	0.29	2.21
25 x 150mm	m	0.19	1.52	1.00	0.38	2.90
38 x 75mm	m	0.15	1.20	0.80	0.30	2.30
38 x 100mm	m	0.20	1.60	1.08	0.40	3.08
50 x 50mm	m	0.14	1.12	0.59	0.26	1.97
50 x 75mm	m	0.20	1.60	0.80	0.36	2.76
50 x 100mm	m	0.26	2.08	1.06	0.47	3.61
75 x 75mm	m	0.28	2.24	1.34	0.54	4.12
75 x 100mm	m	0.34	2.72	1.80	0.68	5.20
75 x 125mm	m	0.42	3.36	2.33	0.85	6.54
Solid strutting						
38 x 100mm	m	0.45	3.60	1.08	0.70	5.38
50 x 100mm	m	0.50	4.00	1.06	0.76	5.82
50 x 125mm	m	0.50	4.00	1.31	0.80	6.11
50 x 150mm	m	0.50	4.00	1.60	0.84	6.44
Herringbone strutting 50 x 50mm to joists, depth						
125mm	m	0.60	4.80	0.95	0.86	6.61
150mm	m	0.60	4.80	1.00	0.87	6.67
175mm	m	0.60	4.80	1.30	0.91	7.01
240mm	m	0.60	4.80	1.40	0.93	7.13
Trimming around rectangular openings, joists size						
50 x 100mm	m	1.50	12.00	0.00	1.80	13.80
50 x 125mm	m	1.65	13.20	0.00	1.98	15.18
50 x 150mm	m	1.80	14.40	0.00	2.16	16.56
75 x 125mm	m	2.12	16.96	0.00	2.54	19.50
75 x 150mm	m	2.35	18.80	0.00	2.82	21.62

All the above materials are untreated. 12.5% should be added for cost of preservative treatment

RATES FOR MEASURED WORK

	Unit	Labour hours	Net labour (£)	Net material (£)	O'heads /profit (£)	Total (£)
Gutters, fascias etc.						
Plywood marine quality in gutters over 300mm wide, thickness						
12mm	m2	1.30	10.40	10.19	3.09	23.68
18mm	m2	1.58	12.64	15.03	4.15	31.82
25mm	m2	1.80	14.40	20.86	5.29	40.55
Plywood marine quality in gutters 150mm wide, thickness						
12mm	m	0.42	3.36	1.69	0.76	5.81
18mm	m	0.50	4.00	2.51	0.98	7.49
25mm	m	0.58	4.64	3.43	1.21	9.28
Plywood marine quality in gutters 300mm wide, thickness						
12mm	m	0.68	5.44	3.40	1.33	10.17
18mm	m	0.76	6.08	5.01	1.66	12.75
25mm	m	0.85	6.80	6.99	2.07	15.86
Softwood bearers, size						
25 x 50mm	m	0.09	0.72	0.33	0.16	1.21
38 x 50mm	m	0.11	0.88	0.53	0.21	1.62
50 x 50mm	m	0.12	0.96	0.59	0.23	1.78
Plywood marine grade to eaves, verges, soffits, fascias, thickness						
12mm						
over 300mm width	m2	1.30	10.40	10.19	3.09	23.68
150mm wide	m	0.42	3.36	1.69	0.76	5.81
225mm wide	m	0.53	4.24	2.54	1.02	7.80
18mm						
over 300mm width	m2	1.58	12.64	15.03	4.15	31.82
150mm wide	m	0.50	4.00	3.51	1.13	8.64
225mm wide	m	0.59	4.72	3.76	1.27	9.75

CARPENTRY/TIMBER

	Unit	Labour hours	Net labour (£)	Net material (£)	O'heads /profit (£)	Total (£)
25mm						
over 300mm width	m2	1.80	14.40	20.86	5.29	40.55
150mm wide	m	0.58	4.64	3.47	1.22	9.33
225mm wide	m	0.68	5.44	5.22	1.60	12.26

	Unit	Labour hours	Net labour (£)	Net material (£)	O'heads /profit (£)	Total (£)

G32 WOODWOOL SLAB DECKING

Labour rate 8.00 per hour

Woodcemair unreinforced woodwool
slabs (type 5B) in standard
lengths, fixed to timber joists,
thickness 50mm (type 500)

1800mm lengths	m2	0.68	5.44	6.07	1.73	13.24
2100mm lengths	m2	0.68	5.44	6.07	1.73	13.24
2400mm lengths	m2	0.68	5.44	6.07	1.73	13.24
2700mmm lengths	m2	0.68	5.44	6.15	1.74	13.33
3000mm lengths	m2	0.68	5.44	6.15	1.74	13.33

Woodcemair unreinforced woodwool
slabs (type 5B) in standard
lengths, fixed to timber joists,
thickness 75mm (type 750)

2100mm lengths	m2	0.78	6.24	9.41	2.35	18.00
2400mm lengths	m2	0.78	6.24	9.68	2.39	18.31
2700mm lengths	m2	0.78	6.24	9.68	2.39	18.31
3000mm lengths	m2	0.78	6.24	9.83	2.41	18.48

Woodcemair unreinforced woodwool
slabs (type 5B) in standard
lengths, fixed to timber joists,
thickness 100mm (type 1000)

3000mm lengths	m2	0.86	6.88	13.53	3.06	23.47
3300mm lengths	m2	0.86	6.88	13.56	3.07	23.51

Woodcelip reinforced woodwool
slabs, in standard lengths, fixed
to timber joists thickness 50mm
(type 503)

1800mm lengths	m2	0.90	7.20	13.80	3.15	24.15
2000mm lengths	m2	0.90	7.20	13.80	3.15	24.15
2100mm lengths	m2	0.90	7.20	13.80	3.15	24.15
2400mm lengths	m2	0.90	7.20	14.50	3.25	24.95
2700mm lengths	m2	0.90	7.20	14.73	3.29	25.22
3000mm lengths	m2	0.90	7.20	14.73	3.29	25.22

WOODWOOL SLAB DECKING

	Unit	Labour hours	Net labour (£)	Net material (£)	O'heads /profit (£)	Total (£)
Woodcelip reinforced woodwool slabs, in standard lengths, fixed to timber joists thickness (75mm) (type 751)						
1800mm lengths	m2	0.98	7.84	20.31	4.22	32.37
2000mm lengths	m2	0.98	7.84	20.31	4.22	32.37
2400mm lengths	m2	0.98	7.84	20.31	4.22	32.37
2700mm lengths	m2	0.98	7.84	20.35	4.23	32.42
3000mm lengths	m2	0.98	7.84	20.35	4.23	32.42
Woodwool reinforced woodwool slabs, in standard lengths, fixed to timber joists thickness 75mm (type 752)						
1800mm lengths	m2	0.98	7.84	20.19	4.20	32.23
2000mm lengths	m2	0.98	7.84	20.19	4.20	32.23
2400mm lengths	m2	0.98	7.84	20.19	4.20	32.23
2700mm lengths	m2	0.98	7.84	20.31	4.22	32.37
3000mm lengths	m2	0.98	7.84	20.31	4.22	32.37
Woodcelip reinforced woodwool slabs, in standard lengths, fixed to timber joists thickness 75mm (type 753)						
2400mm lengths	m2	0.98	7.84	20.14	4.20	32.18
2700mm lengths	m2	0.98	7.84	21.01	4.33	33.18
3000mm lengths	m2	0.98	7.84	21.01	4.33	33.18
3300mm lengths	m2	0.98	7.84	24.64	4.87	37.35
3600mm lengths	m2	0.98	7.84	24.64	4.87	37.35
3900mm lengths	m2	0.98	7.84	24.64	4.87	37.35
Woodcelip reinforced woodwool slabs, in standard lengths, fixed to timber joists thickness 100mm (type 1001)						
3000mm lengths	m2	1.12	8.96	26.53	5.32	40.81
3300mm lengths	m2	1.12	8.96	26.53	5.32	40.81
3600mm lengths	m2	1.12	8.96	27.69	5.50	42.15

RATES FOR MEASURED WORK

	Unit	Labour hours	Net labour (£)	Net material (£)	O'heads /profit (£)	Total (£)
Woodwool reinforced woodwool slabs, in standard lengths, fixed to timber joists, thickness 100mm (type 1003)						
3000mm lengths	m2	1. 12	8. 96	24. 86	5. 07	38. 89
3300mm lengths	m2	1. 12	8. 96	24. 86	5. 07	38. 89
3600mm lengths	m2	1. 12	8. 96	24. 86	5. 07	38. 89
3900mm lengths	m2	1. 12	8. 96	24. 86	0. 85	34. 67
4000mm lengths	m2	1. 12	8. 96	24. 86	5. 07	38. 89
Woodelip reinforced woodwool slabs, in standard lengths, fixed to timber joists, thickness 125mm (type 1252)						
2400mm lengths	m2	1. 15	9. 20	28. 09	5. 59	42. 88
2700mm lengths	m2	1. 15	9. 20	28. 09	5. 59	42. 88
3000mm lengths	m2	1. 15	9. 20	28. 09	5. 59	42. 88

H
CLADDING/COVERING

MATERIAL COSTS

	Unit	Material supply (£)	Material waste (%)	Total (£)
Natural slates				
Welsh blue/grey slates, size				
405 x 205mm	100	126.04	2.5	129.19
405 x 255mm	100	138.95	2.5	142.42
405 x 305mm	100	168.31	2.5	172.52
460 x 255mm	100	200.76	2.5	205.78
510 x 305mm	100	269.25	2.5	275.98
560 x 305mm	100	348.72	2.5	357.44
610 x 305mm	100	469.71	2.5	481.45
Westmorland green slates				
500-300mm long	t	1858.50	5.0	1951.43
300-225mm long	t	1050.00	5.0	1102.50
Fibre cement slates				
Duracem				
500 x 250mm	100	84.96	5.0	89.21
600 x 300mm	100	113.16	5.0	118.82
Eternit 2000				
400 x 240mm	100	62.41	5.0	65.53
500 x 250mm	100	86.72	5.0	91.06
600 x 300mm	100	115.28	5.0	121.04
600 x 600mm	100	230.56	5.0	242.09
Rivendale				
500 x 250mm	100	97.07	5.0	101.92
500 x 400mm	100	93.56	5.0	98.24
600 x 300mm	100	129.15	5.0	135.61
600 x 600mm	100	258.29	5.0	271.20
Reconstructed stone slates				
Marley Monarch	1000	1314.60	5.0	1380.33
Redland Cambrian	1000	1401.49	5.0	1471.56

145

RATES FOR MEASURED WORK

	Unit	Material supply (£)	Material waste (%)	Total (£)
Marley Roof Tiles				
Plain	1000	273. 00	2. 5	279. 83
Ludlow Plus	1000	406. 35	2. 5	416. 51
Mendip	1000	745. 50	2. 5	764. 14
Double Roman	1000	657. 30	2. 5	673. 73
Modern	1000	750. 75	2. 5	769. 52
Monarch	1000	1314. 60	2. 5	1347. 47
Aluminium nails	1kg	4. 73	10. 0	5. 20
Bonnet hips, valleys and angles	100	163. 80	5. 0	171. 99
Segmental ridge	100	213. 15	5. 0	223. 81
Modern ridge	100	236. 25	5. 0	248. 06
Mono ridge - Modern	100	386. 40	5. 0	405. 72
Mono ridge - Segmental	100	386. 40	5. 0	405. 72
Dentil slips				
Wessex	100	9. 45	5. 0	9. 92
Mendip	100	9. 45	5. 0	9. 92
Bold Roll	100	9. 45	5. 0	9. 92
Verge slips	100	9. 45	5. 0	9. 92
Ventilation ridge terminal	each	28. 35	5. 0	29. 77
Soil vent terminal	each	27. 30	5. 0	28. 67
Gas vent ridge	each	44. 10	5. 0	46. 31
Tile clips				
Modern	100	2. 26	5. 0	2. 37
Wessex	100	2. 26	5. 0	2. 37
Bold Roll	100	2. 20	2. 5	2. 26
Plain	100	5. 25	5. 0	5. 51

146

MATERIAL COSTS

	Unit	Material supply (£)	Material waste (%)	Total (£)
Gas vent ridge	each	44. 10	5. 0	46. 31
Hip irons				
4mm	each	1. 26	5. 0	1. 32
6mm	each	1. 37	5. 0	1. 44
Ventilated Dry Ridge System				
Modern/major batten section	each	6. 04	5. 0	6. 34
Segmental ridge union	each	0. 70	5. 0	0. 74
Modern ridge union	each	0. 70	5. 0	0. 74
Modern filler unit	each	0. 30	5. 0	0. 32
Mendip/bold roll batten section	each	6. 04	5. 0	6. 34
Mendip filler unit	each	0. 30	5. 0	0. 32
Wessex filler unit	each	0. 30	5. 0	0. 32
Double roman filler unit	each	0. 30	5. 0	0. 32
GVR adaptor unit	each	1. 98	5. 0	2. 08
Dry ridge setting out gauge	each	3. 36	5. 0	3. 53
Eaves Ventilation System				
strip ventilator (1m)	each	1. 73	5. 0	1. 82
under felt support	each	1. 25	5. 0	1. 31
eaves vent duct	each	0. 68	5. 0	0. 71
Profiled eaves fillers				
Mendip	100	1. 05	5. 0	1. 10
Double Roman	100	12. 07	5. 0	12. 67
Eaves vent clips				
Mendip	100	2. 63	5. 0	2. 76
Modern	100	2. 63	5. 0	2. 76
Ludlow	100	2. 63	5. 0	2. 76
Double Roman	100	2. 63	5. 0	2. 76
Eaves clip				
Modern	100	2. 26	5. 0	2. 37
Mendip	100	2. 26	5. 0	2. 37
Double Roman	100	2. 26	5. 0	2. 37

RATES FOR MEASURED WORK

	Unit	Material supply (£)	Material waste (%)	Total (£)
Verge clip				
Modern	100	19.53	5.0	20.51
Eaves filler unit				
Mendip	100	16.27	5.0	17.08
Interlocking dry verge				
LH verge unit	each	1.89	5.0	1.98
RH verge unit	each	1.89	5.0	1.98
LH stop end unit	each	1.89	5.0	1.98
RH stop end unit	each	1.89	5.0	1.98
segmental ridge end cap	each	1.89	5.0	1.98
modern ridge end cap	each	1.89	5.0	1.98
LH mono-ridge end cap	each	3.78	7.5	4.06
RH mono-ridge end cap	each	3.78	7.5	4.06
Redland tiles				
Clay roof tiling				
Renown	1000	679.11	2.5	696.09
Norfolk pantile	1000	486.15	2.5	498.30
Delta	1000	1114.99	2.5	1142.86
Ornamental	1000	383.42	2.5	393.01
Rosemary red	1000	358.39	2.5	367.35
Half round ridge	100	216.09	5.0	226.89
Third round hip	100	216.09	5.0	226.89
Universal valley trough	100	441.00	5.0	463.05
Dentil slips				
41mm	100	22.05	5.0	23.15
55mm	100	22.05	5.0	23.15
83mm	100	22.05	5.0	23.15
Dryvent ridge system	3m	32.34	2.5	33.15
Angle hip	100	231.53	5.0	243.11
Bonnet hip	100	170.89	5.0	179.43

MATERIAL COSTS

	Unit	Material supply (£)	Material waste (%)	Total (£)
Valley tiles	100	170.89	5.0	179.43
Gas flue ridge terminal	each	45.20	2.5	46.33
Cloaked verge tile	100	188.35	2.5	193.06
Universal angle ridge	100	226.01	2.5	231.66
Delta ridge	100	263.69	2.5	270.28
Delta angle hip	100	242.11	2.5	248.16
Dry verge system	5m	40.36	2.5	41.37
Delta flue ridge terminal	each	56.28	2.5	57.69

Fibre cement sheeting

Corrugated reinforced cement
sheeting - Eternit 2000

profile 3 grey	m2	7.33	5.0	7.70
profile 3 coloured	m2	8.53	5.0	8.96
profile 6 grey	m2	7.51	5.0	7.89
profile 6 coloured	m2	8.65	5.0	9.08

Profile 3 grey fittings

ridge fittings	nr	5.23	5.0	5.49
eaves filler	nr	5.11	5.0	5.37
eaves closure 75mm	nr	5.11	5.0	5.37
apron flashing	nr	5.84	5.0	6.13

Profile 6 grey fittings

ridge fittings	nr	6.42	5.0	6.74
eaves filler	nr	6.70	5.0	7.04
eaves closure 100mm	nr	7.22	5.0	7.58
eaves bend sheet 1525mm (300mm radius)	nr	24.57	5.0	25.80
apron flashing	nr	7.22	5.0	7.58

RATES FOR MEASURED WORK

	Unit	Material supply (£)	Material waste (%)	Total (£)
Profile 3 coloured fittings				
ridge fittings	nr	6. 47	5. 0	6. 79
eaves filler	nr	5. 67	5. 0	5. 95
eaves closure 75mm	nr	5. 95	5. 0	6. 25
apron flashing	nr	6. 51	5. 0	6. 84
Profile 6 coloured fittings				
ridge fittings	nr	8. 45	5. 0	8. 87
eaves filler	nr	7. 77	5. 0	8. 16
eaves closure 100mm	nr	8. 36	5. 0	8. 78
eaves bendsheet 1525mm				
(30mm radius)	nr	26. 98	5. 0	28. 33
apron flashing	nr	8. 36	5. 0	8. 78

Translucent sheeting

Corrugated glass fibre reinforced
translucent sheeting 1.3mm thick

75mm profile	m2	6. 91	5. 0	7. 26
150mm profile	m2	10. 25	5. 0	10. 76

Fibre bitumen sheeting

Nuralite

FX	m2	13. 92	5. 0	14. 62

Beaded cover flashings,
preformed, girth

100mm	m	1. 28	5. 0	1. 34
150mm	m	1. 67	5. 0	1. 75
200mm	m	2. 21	5. 0	2. 32
250mm	m	2. 72	5. 0	2. 86
300mm	m	3. 36	5. 0	3. 53

MATERIAL COSTS

	Unit	Material supply (£)	Material waste (%)	Total (£)
Ridge trays, preformed, length				
250mm	nr	1.06	5.0	1.11
350mm	nr	1.25	5.0	1.31
450mm	nr	1.49	5.0	1.56
Intermediate trays, preformed, length				
250mm	nr	1.23	5.0	1.29
350mm	nr	1.49	5.0	1.56
450mm	nr	1.63	5.0	1.71
Catchment trays, preformed, length				
250mm	nr	1.25	5.0	1.31
350mm	nr	1.63	5.0	1.71
450mm	nr	1.63	5.0	1.71
Catchment closure pieces, preformed, girth				
250mm	nr	1.80	5.0	1.89
350mm	nr	0.91	5.0	0.96
450mm	nr	1.05	5.0	1.10
Soakers, preformed, girth				
150mm	nr	0.56	5.0	0.59
175mm	nr	0.62	5.0	0.65
216mm	nr	0.72	5.0	0.76
250mm	nr	0.82	5.0	0.86
300mm	nr	0.95	5.0	1.00
350mm	nr	1.10	5.0	1.16
432mm	nr	1.66	5.0	1.74
Linings to concrete gutters, preformed, girth				
450mm	m	5.74	5.0	6.03
490mm	m	5.74	5.0	6.03

151

RATES FOR MEASURED WORK

	Unit	Material supply (£)	Material waste (%)	Total (£)
Lead sheet coverings				
Sheet lead to BS1178	t	1102. 50	5. 0	1157. 63

FIBRE CEMENT SHEETING

	Unit	Labour hours	Net labour (£)	Net material (£)	O'heads /profit (£)	Total (£)

H30 FIBRE CEMENT SHEETING

Labour rate 8.00 per hour

Corrugated reinforced cement
sheeting, lapped one corrugation
at sides and 150mm at ends, fixed
with screws and washers to timber
purlins

	Unit	Labour hours	Net labour (£)	Net material (£)	O'heads /profit (£)	Total (£)
profile 3 grey sheets	m2	0.75	6.00	8.49	2.17	16.66
profile 3 coloured sheets	m2	0.75	6.00	9.88	2.38	18.26
profile 6 grey sheets	m2	0.70	5.60	8.70	2.14	16.44
profile 6 coloured sheets	m2	0.70	5.60	10.01	2.34	17.95

Corrugated reinforced cement
sheeting, lapped one corrugation
at sides and 150mm at ends, fixed
with hook bolts and washers to
steel purlins

	Unit	Labour hours	Net labour (£)	Net material (£)	O'heads /profit (£)	Total (£)
profile 3 grey sheets	m2	0.85	6.80	8.49	2.29	17.58
profile 3 coloured sheets	m2	0.85	6.80	9.88	2.50	19.18
profile 6 grey sheets	m2	0.80	6.40	8.70	2.26	17.36
profile 6 coloured sheets	m2	0.80	6.40	10.01	2.46	18.87

Fittings to profile 3 sheets

	Unit	Labour hours	Net labour (£)	Net material (£)	O'heads /profit (£)	Total (£)
ridge fitting	nr	0.35	2.80	6.06	1.33	10.19
eaves filler	nr	0.20	1.60	5.92	1.13	8.65
eaves closure 75mm	nr	0.20	1.60	5.92	1.13	8.65
apron flashing	nr	0.25	2.00	6.73	1.31	10.04

Fittings to profile 6 sheets

	Unit	Labour hours	Net labour (£)	Net material (£)	O'heads /profit (£)	Total (£)
ridge fitting	nr	0.35	2.80	7.43	1.53	11.76
eaves filler	nr	0.20	1.60	7.76	1.40	10.76
eaves closure 100mm	nr	0.20	1.60	7.98	1.44	11.02
eaves bend sheet 1525mm (300mm radius)	nr	0.35	2.80	32.10	5.24	40.14
apron flashing	nr	0.25	2.00	8.36	1.55	11.91

153

RATES FOR MEASURED WORK

	Unit	Labour hours	Net labour (£)	Net material (£)	O'heads /profit (£)	Total (£)

H31 METAL PROFILED SHEETING

Labour rate 8.00 per hour

Aluminium alloy roll formed profiled sheets (Precision Metal Forming Ltd)

Wall cladding, fixed to steelwork with self tapping screws, 0.55mm thick

profile 13.5/3	m2	0.40	3.20	5.69	1.33	10.22
profile 19	m2	0.42	3.36	5.32	1.30	9.98
profile R32	m2	0.45	3.60	6.02	1.44	11.06
profile R35A	m2	0.47	3.76	6.48	1.54	11.78
profile R38A	m2	0.48	3.84	7.69	1.73	13.26
profile R46	m2	0.50	4.00	7.23	1.68	12.91

Wall cladding, fixed to steelwork with self tapping screws, 0.70mm thick

profile 13.5/3	m2	0.40	3.20	6.90	1.52	11.62
profile 19	m2	0.42	3.36	6.47	1.47	11.30
profile R32	m2	0.45	3.60	7.30	1.64	12.54
profile R35A	m2	0.47	3.76	7.90	1.75	13.41
profile R38A	m2	0.48	3.84	7.77	1.74	13.35
profile R46	m2	0.50	4.00	8.01	1.80	13.81
profile R60	m2	0.56	4.48	9.10	2.04	15.62

Wall cladding, fixed to steelwork with self tapping screws, 0.90mm thick

profile 13.5/3	m2	0.40	3.20	8.69	1.78	13.67
profile 19	m2	0.42	3.36	8.18	1.73	13.27
profile R32	m2	0.45	3.60	9.22	1.92	14.74
profile R35A	m2	0.47	3.76	9.94	2.05	15.75
profile R38A	m2	0.48	3.84	9.78	2.04	15.66
profile R46	m2	0.50	4.00	10.09	2.11	16.20
profile R60	m2	0.56	4.48	11.52	2.40	18.40
profile R100	m2	0.60	4.80	13.49	2.74	21.03

154

METAL PROFILED SHEETING

	Unit	Labour hours	Net labour (£)	Net material (£)	O'heads /profit (£)	Total (£)
Wall cladding, fixed to steelwork with self tapping screws, 1.20mm thick						
profile 13.5/3	m2	0.40	3.20	11.59	2.22	17.01
profile 19	m2	0.42	3.36	10.90	2.14	16.40
profile R32	m2	0.45	3.60	12.27	2.38	18.25
profile R35A	m2	0.47	3.76	13.27	2.55	19.58
profile R38A	m2	0.48	3.84	13.06	2.54	19.44
profile R46	m2	0.50	4.00	13.44	2.62	20.06
profile R60	m2	0.56	4.48	15.35	2.97	22.80
profile R100	m2	0.60	4.80	18.03	3.42	26.25
Roof cladding, fixed to steelwork with self tapping screws, 0.55mm thick						
profile 13.5/3	m2	0.35	2.80	5.69	1.27	9.76
profile 19	m2	0.37	2.96	5.32	1.24	9.52
profile R32	m2	0.39	3.12	6.01	1.37	10.50
profile R35A	m2	0.40	3.20	6.48	1.45	11.13
profile R38A	m2	0.42	3.36	7.69	1.66	12.71
profile R46	m2	0.44	3.52	7.23	1.61	12.36
Roof cladding, fixed to steelwork with self tapping screws, 0.70mm thick						
profile 13.5/3	m2	0.35	2.80	6.90	1.46	11.16
profile 19	m2	0.37	2.96	6.47	1.41	10.84
profile R32	m2	0.39	3.12	7.30	1.56	11.98
profile R35A	m2	0.40	3.20	7.90	1.67	12.77
profile R38A	m2	0.42	3.36	7.77	1.67	12.80
profile R46	m2	0.44	3.52	8.22	1.76	13.50
profile R60	m2	0.52	4.16	8.33	1.87	14.36
Roof cladding, fixed to steelwork with self tapping screws, 0.90mm thick						
profile 13.5/3	m2	0.33	2.64	8.70	1.70	13.04
profile 19	m2	0.37	2.96	8.18	1.67	12.81
profile R32	m2	0.39	3.12	9.23	1.85	14.20
profile R35A	m2	0.40	3.20	9.94	1.97	15.11

Roof cladding (cont'd)	Unit	Labour hours	Net labour (£)	Net material (£)	O'heads /profit (£)	Total (£)
profile R38A	m2	0.42	3.36	9.78	1.97	15.11
profile R46	m2	0.44	3.52	10.09	2.04	15.65
profile R60	m2	0.52	4.16	11.52	2.35	18.03
profile R100	m2	0.56	4.48	13.49	2.70	20.67

Roof cladding, fixed to steelwork
with self tapping screws, 1.20mm
thick

profile 13.5/3	m2	0.35	2.80	11.59	2.16	16.55
profile 19	m2	0.37	2.96	10.90	2.08	15.94
profile R32	m2	0.39	3.12	12.27	2.31	17.70
profile R35A	m2	0.40	3.20	13.27	2.47	18.94
profile R38A	m2	0.42	3.36	13.06	2.46	18.88
profile R46	m2	0.44	3.52	13.44	2.54	19.50
profile R60	m2	0.52	4.16	15.35	2.93	22.44
profile R100	m2	0.56	4.48	18.03	3.38	25.89

Aluminium alloy pressed sheets

Wall cladding, fixed to steelwork
with self tapping screws, 0.55mm
thick

profile PR8	m2	0.42	3.36	5.51	1.33	10.20
profile PM13	m2	0.43	3.44	5.89	1.40	10.73
profile PL19	m2	0.44	3.52	6.10	1.44	11.06
profile PG22	m2	0.45	3.60	6.46	1.51	11.57
profile PS47	m2	0.56	4.48	8.00	1.87	14.35

Wall cladding, fixed to steelwork
with self tapping screws, 0.70mm
thick

profile PR8	m2	0.42	3.36	6.69	1.51	11.56
profile PM13	m2	0.43	3.44	7.15	1.59	12.18
profile PL19	m2	0.44	3.52	7.40	1.64	12.56
profile PG22	m2	0.45	3.60	7.85	1.72	13.17
profile PS47	m2	0.56	4.48	9.70	2.13	16.31

METAL PROFILED SHEETING

	Unit	Labour hours	Net labour (£)	Net material (£)	O'heads /profit (£)	Total (£)
Wall cladding, fixed to steelwork with self tapping screws, 0.90mm thick						
profile PR8	m2	0.42	3.36	8.47	1.77	13.60
profile PM13	m2	0.43	3.44	9.05	1.87	14.36
profile PL19	m2	0.44	3.52	9.35	1.93	14.80
profile PG22	m2	0.45	3.60	9.93	2.03	15.56
profile PS47	m2	0.56	4.48	12.29	2.52	19.29
Wall cladding, fixed to steelwork with self tapping screws, 1.20mm thick						
profile PR8	m2	0.42	3.36	11.28	2.20	16.84
profile PL19	m2	0.44	3.52	12.46	2.40	18.38
profile PG22	m2	0.45	3.60	13.21	2.52	19.33
profile PS47	m2	0.56	4.48	16.35	3.12	23.95
Roof cladding, fixed to steelwork with self tapping screws, 0.55mm thick						
profile PR8	m2	0.37	2.96	5.51	1.27	9.74
profile PM13	m2	0.39	3.12	5.89	1.35	10.36
profile PL19	m2	0.41	3.28	6.10	1.41	10.79
profile PG22	m2	0.43	3.44	6.46	1.48	11.38
profile PS47	m2	0.44	3.52	8.00	1.73	13.25
Roof cladding, fixed to steelwork with self tapping screws, 0.70mm thick						
profile PR8	m2	0.39	3.12	6.69	1.47	11.28
profile PM13	m2	0.39	3.12	7.15	1.54	11.81
profile PL19	m2	0.41	3.28	7.40	1.60	12.28
profile PG22	m2	0.43	3.44	7.85	1.69	12.98
profile PS47	m2	0.44	3.52	9.70	1.98	15.20

157

Roof cladding (cont'd)	Unit	Labour hours	Net labour (£)	Net material (£)	O'heads /profit (£)	Total (£)
Roof cladding, fixed to steelwork with self tapping screws, 0.90mm thick						
profile PR8	m2	0.37	2.96	8.48	1.72	13.16
profile PM13	m2	0.39	3.12	9.05	1.83	14.00
profile PL19	m2	0.41	3.28	9.35	1.89	14.52
profile PG22	m2	0.43	3.44	9.93	2.01	15.38
profile PS47	m2	0.44	3.52	12.29	2.37	18.18
Roof cladding, fixed to steelwork with self tapping screws, 1.20mm thick						
profile PR8	m2	0.37	2.96	11.28	2.14	16.38
profile PL19	m2	0.41	3.28	12.45	2.36	18.09
profile PG22	m2	0.43	3.44	13.21	2.50	19.15
profile PS47	m2	0.44	3.52	16.35	2.98	22.85
Alugalve roll formed profiled sheets						
Roof cladding, fixed to steelwork with self tapping screws, 0.70mm thick						
profile 13.5/3	m2	0.35	2.80	6.48	1.39	10.67
profile 19	m2	0.37	2.96	6.11	1.36	10.43
profile 32(s)(v)	m2	0.39	3.12	6.87	1.50	11.49
profile 35A	m2	0.40	3.20	7.46	1.60	12.26
profile 38A	m2	0.42	3.36	7.33	1.60	12.29
profile 46	m2	0.44	3.52	7.56	1.66	12.74
profile 60	m2	0.56	4.48	8.62	1.96	15.06
profile 100	m2	0.60	4.80	10.00	2.22	17.02
Roof cladding, fixed to steelwork with self tapping screws, 0.70mm thick						
profile PR8	m2	0.37	2.96	6.31	1.39	10.66
profile PM13	m2	0.39	3.12	6.75	1.48	11.35
profile PL19	m2	0.41	3.28	6.98	1.54	11.80
profile PG22	m2	0.43	3.44	7.40	1.63	12.47
profile PS47	m2	0.44	3.52	9.16	1.90	14.58

METAL PROFILED SHEETING

	Unit	Labour hours	Net labour (£)	Net material (£)	O'heads /profit (£)	Total (£)
Colour coated galvanized steel roll formed profiled sheets, Colourcoat HP 200 Leathergrain embossed PVC Plastisol one side, standard light grey backing coat						
Wall cladding, fixed to steelwork with self tapping screws, 0.55mm thick						
profile 13.5/3	m2	0.40	3.20	8.79	1.80	13.79
profile 19	m2	0.42	3.36	8.24	1.74	13.34
profile 32(s)(v)	m2	0.45	3.60	9.28	1.93	14.81
profile 35A	m2	0.47	3.76	9.77	2.03	15.56
profile 38A	m2	0.48	3.84	9.63	2.02	15.49
profile 46	m2	0.50	4.00	10.14	2.12	16.26
profile 60	m2	0.56	4.48	11.56	2.41	18.45
Roof cladding, fixed to steelwork with self tapping screws, 0.55mm thick						
profile 13.5/3	m2	0.35	2.80	8.79	1.74	13.33
profile 19	m2	0.37	2.96	8.24	1.68	12.88
profile 32(s)(v)	m2	0.39	3.12	9.28	1.86	14.26
profile 35A	m2	0.40	3.20	9.77	1.95	14.92
profile 38A	m2	0.42	3.36	9.63	1.95	14.94
profile 46	m2	0.44	3.52	10.14	2.05	15.71
profile 60	m2	0.52	4.16	11.56	2.36	18.08

159

RATES FOR MEASURED WORK

	Unit	Labour hours	Net labour (£)	Net material (£)	O'heads /profit (£)	Total (£)

H41 TRANSLUCENT SHEETING

Labour rate 8.00 per hour

Corrolux vinyl sheets to BS4203
translucent sheeting 1.3mm thick,
lapped one corrugation at sides
and 150mm at ends fixed with
screws and washers to timber
purlins

	Unit	Labour hours	Net labour	Net material	O'heads /profit	Total
profile 3 sheets	m2	0.75	6.00	7.62	0.34	13.96
profile 6 sheets	m2	0.70	5.60	11.30	0.42	17.32

Corrolux vinyl sheets to BS4203
translucent sheeting 1.3mm thick,
lapped one corrugation at sides
and 150mm at ends fixed with hook
bolts and washers to steel purlins

	Unit	Labour hours	Net labour	Net material	O'heads /profit	Total
profile 3 sheets	m2	0.85	6.80	7.62	0.36	14.78
profile 6 sheets	m2	0.80	6.40	11.30	0.44	18.14

CLAY/CONCRETE ROOF TILING

	Unit	Labour hours	Net labour (£)	Net material (£)	O'heads /profit (£)	Total (£)

H60 CLAY/CONCRETE ROOF TILING

Labour rate 8.00 per hour

Marley Plain granuled or smooth finish tiles size 267 x 165mm, 65mm lap, 35 degrees pitch, type 1F reinforced underlay

Battens size 38 x 19mm

	Unit	Labour hours	Net labour (£)	Net material (£)	O'heads /profit (£)	Total (£)
gauge 100mm	m2	1.73	13.84	20.13	5.10	39.07
gauge 95mm	m2	1.75	14.00	21.37	5.31	40.68
gauge 90mm	m2	1.78	14.24	22.67	5.54	42.45

Battens size 38 x 25mm

	Unit	Labour hours	Net labour (£)	Net material (£)	O'heads /profit (£)	Total (£)
gauge 100mm	m2	1.83	14.64	20.55	5.28	40.47
gauge 95mm	m2	1.85	14.80	21.59	5.46	41.85
gauge 90mm	m2	1.88	15.04	22.91	5.69	43.64

Extra for

	Unit	Labour hours	Net labour (£)	Net material (£)	O'heads /profit (£)	Total (£)
nailing every tile with aluminium nails	m2	0.30	2.40	0.21	0.39	3.00
interlocking dry verge system	m	0.20	1.60	7.43	1.35	10.38
verge, 150mm wide plain tile undercloak	m	0.20	1.60	1.02	0.39	3.01
double course at eaves	m	0.35	2.80	1.66	0.67	5.13
segmental ridge tile	m	0.42	3.36	6.33	1.45	11.14
segmental monoridge tiles	m	0.62	4.96	10.06	2.25	17.27
valley trough tiles	m	0.62	4.96	6.29	1.69	12.94
segmental hip tiles	m	0.62	4.96	6.29	1.69	12.94
bonnet hip tiles	m	0.80	6.40	6.29	1.90	14.59
Marley eaves vent system	m	0.40	3.20	10.27	2.02	15.49
ventilated ridge terminal	nr	0.60	4.80	29.77	5.19	39.76
gas vent terminal	nr	0.60	4.80	46.31	7.67	58.78
soil vent terminal	nr	0.60	4.80	28.67	5.02	38.49
cutting	m	0.20	1.60	0.00	0.24	1.84
holes for pipes	nr	0.35	2.80	0.00	0.42	3.22

RATES FOR MEASURED WORK

	Unit	Labour hours	Net labour (£)	Net material (£)	O'heads /profit (£)	Total (£)
Marley Ludlow Plus smooth finish tiles size 387 x 229mm, battens size 38 x 25mm, type 1F reinforced underlay						
75mm lap, pitch 25 to 44 degrees	m2	0.93	7.44	9.01	0.41	16.86
100mm lap, pitch 22 to 44 degrees	m2	1.02	8.16	9.08	0.43	17.67
Extra for						
nailing every tile with aluminium nails	m2	0.09	0.72	0.36	0.16	1.24
interlocking dry verge system	m	0.20	1.60	7.43	1.35	10.38
verge, 150mm wide plain tile undercloak	m	0.20	1.60	1.02	0.39	3.01
segmental ridge tiles	m	0.42	3.36	6.33	1.45	11.14
segmental monoridge tiles	m	0.62	4.96	10.06	2.25	17.27
dry ridge system	m	0.50	4.00	6.24	1.54	11.78
valley trough tiles	m	0.62	4.96	6.29	1.69	12.94
segmental hip tiles	m	0.62	4.96	6.29	1.69	12.94
Marley eaves vent system	m	0.40	3.20	10.27	2.02	15.49
ventilated ridge terminal	nr	0.60	4.80	29.77	5.19	39.76
gas vent terminal	nr	0.60	4.80	44.31	7.37	56.48
soil vent terminal	nr	0.60	4.80	28.67	5.02	38.49
cutting	m	0.20	1.60	0.00	0.04	1.64
holes for pipes	nr	0.35	2.80	0.00	0.42	3.22
Marley Modern smooth finish tiles size 420 x 330mm, battens size 38 x 25mm, type 1F reinforced underlay						
75mm lap, pitch 22.5 to 44 degrees	m2	0.82	6.56	9.37	2.39	18.32
Extra for						
nailing every tile with aluminium nails	m2	0.05	0.40	0.21	0.09	0.70
verge, 150mm wide plain tile undercloak	m	0.20	1.60	1.02	0.39	3.01
dry verge system with white PVC interlocking units	m	0.20	1.60	4.73	0.95	7.28

162

CLAY/CONCRETE ROOF TILING

	Unit	Labour hours	Net labour (£)	Net material (£)	O'heads /profit (£)	Total (£)
Modern ridge tiles	m	0.62	4.96	6.87	1.77	13.60
Modern monoridge	m	0.62	4.96	10.06	2.25	17.27
dry ridge system	m	0.50	4.00	6.24	1.54	11.78
Modern hip tiles	m	0.62	4.96	6.87	1.77	13.60
Marley eaves vent system	m	0.40	3.20	10.27	2.02	15.49
ventilated ridge terminal	nr	0.60	4.80	29.77	5.19	39.76
gas vent terminal	nr	0.60	4.80	46.31	7.67	58.78
soil vent terminal	nr	0.60	4.80	28.67	5.02	38.49
cutting	m	0.20	1.60	0.00	0.24	1.84
holes for pipes	nr	0.35	2.80	0.00	0.42	3.22

Redland Renown granular faced or through coloured tiles size 418 x 330mm, 75mm lap, 343mm gauge, pitch 30 to 40 degrees, type 1F reinforced underlay

	Unit	Labour hours	Net labour (£)	Net material (£)	O'heads /profit (£)	Total (£)
Battens size 38 x 22mm	m2	0.70	5.60	8.35	2.09	16.04
Battens size 38 x 25mm	m2	0.74	5.92	8.41	2.15	16.48

Extra for

	Unit	Labour hours	Net labour (£)	Net material (£)	O'heads /profit (£)	Total (£)
nailing every tile with two aluminium nails	m2	0.09	0.72	0.41	0.17	1.30
cloaked verge tile	m	0.20	1.60	7.17	1.32	10.09
half round ridge or hip tile	m	0.55	4.40	7.17	1.74	13.31
third round ridge tile	nr	0.55	4.40	4.96	1.40	10.76
gas flue ridge terminal	nr	0.70	5.60	45.20	7.62	58.42
Dryvent ridge	m	0.75	6.00	11.32	2.60	19.92
third round hip tile	m	0.55	4.40	5.26	1.45	11.11
Universal valley trough	m	0.30	2.40	12.35	2.21	16.96
cutting	m	0.20	1.60	0.00	0.24	1.84
holes for pipes	nr	0.35	2.80	0.00	0.42	3.22

RATES FOR MEASURED RATES

	Unit	Labour hours	Net labour (£)	Net material (£)	O'heads /profit (£)	Total (£)
Redland Norfolk pantile through coloured tiles size 381 x 227mm, 100mm lap, 306mm gauge, pitch 22.5 to 25 degrees, type 1F reinforced underlay						
Battens size 38 x 22mm	m2	0.80	6.40	9.79	2.43	18.62
Battens size 38 x 25mm	m2	0.89	7.12	10.36	2.62	20.10
Extra for						
nailing every tile with two aluminium nails	m2	0.16	1.28	0.68	0.29	2.25
half round ridge or hip tile	m	0.55	4.40	4.85	1.39	10.64
dryvent ridge	m	0.75	6.00	11.32	2.60	19.92
gas flue ridge terminal	nr	0.70	5.60	45.20	7.62	58.42
third round hip tile	m	0.55	4.40	4.96	1.40	10.76
third round hip and dentil slips 83mm wide	m	0.60	4.80	5.26	1.51	11.57
universal valley trough	m	0.30	2.40	12.35	2.21	16.96
cutting	m	0.20	1.60	0.00	0.04	1.64
holes for pipes	nr	0.35	2.80	0.00	0.42	3.22

FIBRE CEMENT SLATING

	Unit	Labour hours	Net labour (£)	Net material (£)	O'heads /profit (£)	Total (£)
H61 FIBRE CEMENT SLATING						
Labour rate 8.00 per hour						
'Duracem' non-asbestos fibre cement slates size 500 x 250mm, pitch over 40 degrees, 38 x 25mm softwood battens, type 1F reinforced underlay						
lap 70mm, gauge 215mm	m2	1.06	8.48	19.42	4.19	32.09
'Duracem' non-asbestos fibre cement slates size 500 x 250mm, pitch over 27.5 degrees, 38 x 25mm softwood battens, type 1F reinforced underlay						
lap 80mm, gauge 210mm	m2	1.06	8.48	19.05	4.13	31.66
'Duracem' non-asbestos fibre cement slates size 500 x 250mm, pitch 25 to 30 degrees, 38 x 25mm softwood battens, type 1F reinforced underlay						
lap 106mm, gauge 197mm	m2	1.09	8.72	19.90	4.29	32.91
'Duracem' non-asbestos fibre cement slates size 600 x 300mm, pitch over 30 degrees, 38 x 25mm softwood battens, type 1F reinforced underlay						
lap 80mm, gauge 260mm	m2	0.90	7.20	16.05	3.49	26.74
'Duracem' non-asbestos fibre cement slates size 600 x 300mm, pitch over 20 degrees, 38 x 25mm softwood battens, type 1F reinforced underlay						
lap 100mm, gauge 250mm	m2	0.91	7.28	17.72	3.75	28.75

165

	Unit	Labour hours	Net labour (£)	Net material (£)	O'heads /profit (£)	Total (£)
'Duracem' non-asbestos fibre cement slates size 600 x 300mm, pitch 20 to 25 degrees, 38 x 25mm softwood battens, type 1F reinforced underlay						
lap 106mm, gauge 247mm	m2	0.91	7.28	17.97	0.63	25.88
'Eternit 2000' non-asbestos fibre cement slates size 400 x 200mm, pitch over 40 degrees, 38 x 25mm softwood battens, type 1F reinforced underlay						
lap 70mm, gauge 165mm	m2	1.25	10.00	22.00	4.80	36.80
'Eternit 2000' non-asbestos fibre cement slates size 400 x 200mm, pitch 40 to 45 degrees, 38 x 25mm softwood battens, type 1F reinforced underlay						
lap 90mm, gauge 155mm	m2	1.28	10.24	23.58	5.07	38.89
'Eternit 2000' non-asbestos fibre cement slates size 500 x 250mm, pitch over 25 degrees, 38 x 25mm softwood battens, type 1F reinforced underlay						
lap 90mm, gauge 205mm	m2	1.07	8.56	19.79	4.25	32.60
'Eternit 2000' non-asbestos fibre cement slates size 600 x 300mm, pitch over 20 degrees, 38 x 25mm softwood battens, type 1F reinforced underlay						
lap 100mm, gauge 250mm	m2	0.91	7.28	18.02	3.79	29.09

NATURAL SLATING

	Unit	Labour hours	Net labour (£)	Net material (£)	O'heads /profit (£)	Total (£)
H62 NATURAL SLATING						
Labour rate 8.00 per hour						
Blue/grey slates size 405 x 205mm, 75mm lap, 50 x 25mm softwood battens, type 1F reinforced underlay						
Sloping	m2	1.64	13.12	39.23	7.85	60.20
Vertical	m2	1.75	14.00	39.23	7.98	61.21
Mansard	m2	1.75	14.00	39.23	7.98	61.21
Extra for						
double eaves course	m	0.50	4.00	5.95	1.49	11.44
single verge undercloak course	m	0.72	5.76	6.17	1.79	13.72
angled ridge or hip tiles	m	0.70	5.60	13.59	2.88	22.07
mitred hips, cutting both sides	m	0.70	5.60	14.29	2.98	22.87
cutting	m	0.60	4.80	0.00	0.72	5.52
hole for small pipes	nr	0.40	3.20	0.00	0.48	3.68
fix only lead soakers	nr	0.45	3.60	0.00	0.54	4.14
Blue/grey slates size 405 x 255mm, 75mm lap, 50 x 25mm softwood battens, type 1F reinforced underlay						
Sloping	m2	1.35	10.80	41.38	7.83	60.01
Vertical	m2	1.45	11.60	41.38	7.95	60.93
Mansard	m2	1.45	11.60	41.38	7.95	60.93
Extra for						
double eaves course	m	0.50	4.00	6.28	1.54	11.82
single verge undercloak course	m	0.72	5.76	8.00	2.06	15.82
angled ridge or hip tiles	m	0.70	5.60	13.59	2.88	22.07
mitred hips, cutting both sides	m	0.70	5.60	14.29	2.98	22.87

Slating (cont'd)	Unit	Labour hours	Net labour (£)	Net material (£)	O'heads /profit (£)	Total (£)
cutting	m	0.60	4.80	0.00	0.72	5.52
hole for small pipes	nr	0.40	3.20	0.00	0.48	3.68
fix only lead soakers	nr	0.45	3.60	0.00	0.54	4.14

Blue/grey slates size 405 x 305mm, 75mm lap. 50 x 25mm softwood battens, type 1F reinforced underlay

Sloping	m2	1.23	9.84	39.78	7.44	57.06
Vertical	m2	1.33	10.64	39.78	7.56	57.98
Mansard	m2	1.33	10.64	39.78	7.56	57.98

Extra for

double eaves course	m	0.50	4.00	6.35	1.55	11.90
single verge undercloak course	m	0.72	5.76	9.69	2.32	17.77
angled ridge or hip tiles	m	0.70	5.60	13.59	2.88	22.07
mitred hips, cutting both sides	m	0.70	5.60	14.29	2.98	22.87
cutting	m	0.60	4.80	0.00	0.72	5.52
hole for small pipes	nr	0.40	3.20	0.00	0.48	3.68
fix only lead soakers	nr	0.45	3.60	0.00	0.54	4.14

Blue/grey slates size 460 x 255mm, 75mm lap, 50 x 25mm softwood battens, type 1F reinforced underlay

Sloping	m2	1.24	9.92	42.57	7.87	60.36
Vertical	m2	1.35	10.80	42.57	8.01	61.38
Mansard	m2	1.35	10.80	42.57	8.01	61.38

Extra for

double eaves course	m	0.50	4.00	6.36	1.55	11.91
single verge undercloak course	m	0.72	5.76	8.44	2.13	16.33
angled ridge or hip tiles	m	0.07	0.56	13.59	2.12	16.27
mitred hips, cutting both sides	m	0.70	5.60	14.29	2.98	22.87

NATURAL SLATING

	Unit	Labour hours	Net labour (£)	Net material (£)	O'heads /profit (£)	Total (£)
cutting	m	0.60	4.80	0.00	0.72	5.52
hole for small pipes	nr	0.40	3.20	0.00	0.48	3.68
fix only lead soakers	nr	0.45	3.60	0.00	0.54	4.14

Blue/grey slates size 510 x 305mm, 75mm lap, 50 x 25mm softwood battens, type 1F reinforced underlay

Sloping	m2	1.01	8.08	49.04	8.57	65.69
Vertical	m2	1.10	8.80	49.04	8.68	66.52
Mansard	m2	1.10	8.80	49.04	8.68	66.52

Extra for

double eaves course	m	0.50	4.00	10.61	2.19	16.80
single verge undercloak course	m	0.72	5.76	12.39	2.72	20.87
angled ridge or hip tiles	m	0.70	5.60	13.59	2.88	22.07
mitred hips, cutting both sides	m	0.70	5.60	14.29	2.98	22.87
cutting	m	0.60	4.80	0.00	0.72	5.52
hole for small pipes	nr	0.40	3.20	0.00	0.48	3.68
fix only lead soakers	nr	0.45	3.60	0.00	0.54	4.14

Blue/grey slates size 560 x 305mm, 75mm lap, 50 x 25mm softwood battens, type 1F reinforced underlay

Sloping	m2	0.96	7.68	58.72	9.96	76.36

Extra for

double eaves course	m	0.50	4.00	13.03	2.55	19.58
single verge undercloak course	m	0.72	5.76	16.04	3.27	25.07
angled ridge or hip tiles	m	0.70	5.60	13.59	2.88	22.07
mitred hips, cutting both sides	m	0.70	5.60	14.29	2.98	22.87
cutting	m	0.60	4.80	0.00	0.72	5.52
hole for small pipes	nr	0.40	3.20	0.00	0.48	3.68
fix only lead soakers	nr	0.45	3.60	0.00	0.54	4.14

RATES FOR MEASURED WORK

	Unit	Labour hours	Net labour (£)	Net material (£)	O'heads /profit (£)	Total (£)

H63 RECONSTRUCTED STONE SLATING

Labour rate 8.00 per hour

Marley Monarch interlocking slate 38 x 25mm, type 1F reinforced underlay

	Unit	Labour hours	Net labour (£)	Net material (£)	O'heads/profit (£)	Total (£)
75mm lap, pitch 25 to 90 degrees	m2	1.24	9.92	20.27	4.53	34.72
100mm lap, pitch 25 to 90 degrees	m2	1.26	10.08	21.98	0.80	32.86

Extra for

nailing every tile with aluminium nails	m2	0.09	0.72	0.21	0.14	1.07
dry verge system with white PVC interlocking units	m	0.20	1.60	4.73	0.95	7.28
modern ridge tiles	m	0.62	4.96	6.87	1.77	13.60
modern monoridge	m	0.62	4.96	10.06	2.25	17.27
dry ridge system	m	0.50	4.00	6.24	1.54	11.78
modern hip tiles	m	0.62	4.96	6.87	1.77	13.60
ventilated ridge terminal	nr	0.60	4.80	29.77	5.19	39.76
gas vent terminal	nr	0.60	4.80	46.31	7.67	58.78
soil vent terminal	nr	0.60	4.80	28.67	5.02	38.49
cutting	m	0.20	1.60	0.00	0.24	1.84
holes for pipes	nr	0.35	2.80	0.00	0.42	3.22

Redland Cambrian through coloured slates interlocking size 300 x 336mm, pitch 25 to 69 degrees, batten size 38 x 25mm, type 1F reinforcing underlay

	Unit	Labour hours	Net labour (£)	Net material (£)	O'heads/profit (£)	Total (£)
50mm lap, 250mm gauge	m2	0.90	7.20	21.37	4.29	32.86
90mm lap, 210mm gauge	m2	1.00	8.00	25.52	5.03	38.55

Extra for

nailing every tile with two aluminium nails	m2	0.60	4.80	0.41	0.78	5.99
half round ridge tile	m	0.55	4.40	4.96	1.40	10.76
dryvent ridge	m	0.75	6.00	11.32	2.60	19.92

RECONSTRUCTED STONE SLATING

	Unit	Labour hours	Net labour (£)	Net material (£)	O'heads /profit (£)	Total (£)
universal angle ridge	m	0.55	4.40	5.78	1.53	11.71
gas flue ridge terminal	nr	0.70	5.60	46.34	7.79	59.73
universal angle ridge tile as hip	m	0.60	4.80	5.78	1.59	12.17
cutting	m	0.20	1.60	0.00	0.24	1.84
holes for pipes	nr	0.35	2.80	0.00	0.42	3.22

171

	Unit	Labour hours	Net labour (£)	Net material (£)	O'heads /profit (£)	Total (£)

H71 LEAD SHEET COVERINGS

Labour rate 8.00 per hour

Roof coverings, milled sheet lead to BS1178

Flat roofing, pitch less than 10 degrees to the horizontal

code 5	m2	4. 20	33. 60	30. 82	9. 66	74. 08
code 6	m2	4. 40	35. 20	36. 52	10. 76	82. 48

Dormers, pitch less than 10 degrees to the horizontal

code 5	m2	5. 00	40. 00	32. 40	10. 86	83. 26
code 6	m2	5. 25	42. 00	38. 35	12. 05	92. 40

Flashings, horizontal, girth 150mm

code 5	m	0. 45	3. 60	4. 86	1. 27	9. 73

Flashings, horizontal, girth 200mm

code 5	m	0. 60	4. 80	6. 48	1. 69	12. 97

Flashings, horizontal, girth 300mm

code 5	m	0. 55	4. 40	9. 72	2. 12	16. 24

Flashings, sloping, girth 150mm

code 5	m	0. 45	3. 60	4. 86	1. 27	9. 73

Flashings, sloping, girth 200mm

code 5	m	0. 60	4. 80	6. 48	1. 69	12. 97

Flashings, sloping, girth 300mm

code 5	m	0. 85	6. 80	9. 72	2. 48	19. 00

LEAD SHEET COVERINGS

	Unit	Labour hours	Net labour (£)	Net material (£)	O'heads /profit (£)	Total (£)
Aprons, horizontal, girth 200mm						
code 5	m	1.60	12.80	6.48	2.89	22.17
Aprons, horizontal, girth 300mm						
code 5	m	0.90	7.20	9.72	2.54	19.46
Aprons, horizontal, girth 400mm						
code 5	m	1.20	9.60	12.96	3.38	25.94
Sills, horizontal, girth 200mm						
code 5	m	0.60	4.80	6.48	1.69	12.97
Sills, horizontal, girth 300mm						
code 5	m	0.90	7.20	9.72	2.54	19.46
Sills, horizontal, girth 400mm						
code 5	m	1.20	9.60	12.96	3.38	25.94
Hips, sloping, girth 200mm						
code 5	m	0.75	6.00	6.48	1.87	14.35
Hips, sloping, girth 300mm						
code 5	m	1.00	8.00	9.72	2.66	20.38
Hips, sloping, girth 400mm						
code 5	m	1.30	10.40	12.96	3.50	26.86
Kerbs, horizontal, girth 300mm						
code 5	m	0.90	7.20	9.72	2.54	19.46
Kerbs, horizontal, girth 400mm						
code 5	m	1.20	9.60	13.17	3.42	26.19

RATES FOR MEASURED WORK

Leadwork (cont'd)	Unit	Labour hours	Net labour (£)	Net material (£)	O'heads /profit (£)	Total (£)
Valleys, sloping, girth 400mm						
code 5	m	1. 20	9. 60	12. 96	3. 38	25. 94
Valleys, sloping, girth 600mm						
code 5	m	1. 40	11. 20	16. 20	4. 11	31. 51
Valleys, sloping, girth 800mm						
code 5	m	1. 60	12. 80	25. 92	5. 81	44. 53
Gutters, sloping, girth 600mm						
code 5	m	1. 40	11. 20	16. 20	4. 11	31. 51
code 6	m	1. 60	12. 80	19. 17	4. 80	36. 77
Gutters, sloping, girth 800mm						
code 5	m	1. 60	12. 80	25. 92	5. 81	44. 53
code 6	m	1. 80	14. 40	30. 68	6. 76	51. 84
Slates, size 400 x 400mm with 200mm high collar, 100mm diameter						
code 5	nr	1. 60	12. 80	10. 37	0. 58	23. 75
Slates, size 400 x 400mm with 200mm high collar, 150mm diameter						
code 6	nr	1. 80	14. 40	12. 47	4. 03	30. 90

J

WATERPROOFING

The Channel Tunnel Story

Graham Anderson and Ben Roskrow

- unique 'inside' coverage provides a vivid picture of the huge project

- draws on extensive *Construction News* coverage over eight years

- includes colour photographs of the project and of the key people involved, from the board room to the workers at the tunnel face

The Channel Tunnel is a huge construction project, employing over 14,000 people at peak, and costing over £11 billion of private money. It has succeeded in spite of great financial, political and techncial difficulties, and a fundamentally flawed contract. This book tells the story of the project, based on the coverage in *Construction News* and with commentary taken from recent interviews with key project sources.

Contents: Preface. Introduction. Key dates. Key facts. The end of a dream. A fait accompli? The race for the mandate. Sherwood's surprise. Concessions and treaties. Money and politics. Planning and designs. Morton takes charge. High risk, high return. The contractors start work. All change at TML. Tragedy strikes. The £1 billion row. Building the Channel Tunnel. King of the tunnels. TML's £1.4 billion bill. History man. More talks but no trains. Brussels backs Eurotunnel. Bombardier's bombshell. Hail to the chief. Site warfare breaks out. The slow train to France. Epilogue. Index.

May 1994: 234x156: 240pp, 43 halftone illus
Paperback: 0-419-19620-X: £14.99

For further information and to order please contact: The Promotion Dept.,
E & F N Spon, 2-6 Boundary Row, London SE1 8HN
Tel: 071 865 0066 Fax: 071 522 9623

E & F N Spon
An imprint of Chapman & Hall

MATERIAL COSTS

		Unit	Material supply (£)	Material waste (%)	Total (£)
MATERIAL COSTS					
Synthaprufe		5l	9. 07	5. 0	9. 52
Built up felt roof coverings					
Built up roofing to BS747					
Felt type 1B					
14kg/10m2	10m x 1m roll	nr	9. 49	5. 0	9. 96
Felt type 1B					
18kg/10m2	10m x 1m roll	nr	12. 19	5. 0	12. 80
Felt type 1B					
25kg/10m2	10m x 1m roll	nr	17. 73	5. 0	18. 62
Felt type 1E					
38kg/10m2	10m x 1m roll	nr	26. 66	5. 0	27. 99
Felt type 3B					
18kg/10m2	20m x 1m roll	nr	27. 26	5. 0	28. 62
Felt type 3E					
28kg/10m2	10m x 1m roll	nr	20. 42	5. 0	21. 44
Felt type 3G					
28kg/10m2	10m x 1m roll	nr	24. 21	5. 0	25. 42
Felt type 5U					
29kg/10m2	10m x 1m roll	nr	59. 39	5. 0	62. 36
Felt type 5B					
34kg/10m2	10m x 1m roll	nr	49. 85	5. 0	52. 34

	Unit	Material supply (£)	Material waste (%)	Total (£)
Felt type 5E				
38kg/10m2 10m x 1m roll	nr	57. 35	5. 0	60. 22
Felt type Elastomeric				
40kg/10m2 20m x 1m roll	nr	62. 93	5. 0	66. 08
Felt type Elastomeric				
32kg/10m2 10m x 1m roll	nr	43. 52	5. 0	45. 70
Primo	25l	41. 85	10. 0	46. 04
Adhesive	25l	33. 47	10. 0	36. 82
Felt type Euroroof Elastomeric				
G.32	10m2	43. 52	5. 0	45. 70
P.56	10m2	62. 93	5. 0	66. 08

	Unit	Labour hours	Net labour (£)	Net material (£)	O'heads /profit (£)	Total (£)

J20 ASPHALT TANKING

The following are specialist
sub-contractors' prices and
include 15% overheads and profit

Damp-proofing and tanking

20mm two coat mastic asphalt to
BS1097 over 300mm wide

	Unit	Labour hours	Net labour (£)	Net material (£)	O'heads /profit (£)	Total (£)
flat						
sloping 10 to 45 degrees	m2	0. 00	0. 00	23. 11	0. 00	23. 11
sloping 46 to 90 degrees	m2	0. 00	0. 00	26. 13	0. 00	26. 13
vertical	m2	0. 00	0. 00	26. 13	0. 00	26. 13

20mm two coat mastic asphalt to
BS1097, not exceeding 150mm wide

flat	m2	0. 00	0. 00	40. 22	0. 00	40. 22
sloping 10 to 45 degrees	m2	0. 00	0. 00	46. 25	0. 00	46. 25
sloping 46 to 90 degrees	m2	0. 00	0. 00	52. 29	0. 00	52. 29
vertical	m2	0. 00	0. 00	52. 29	0. 00	52. 29

20mm two coat mastic asphalt to
BS1097, 150 to 225mm wide

flat	m2	0. 00	0. 00	35. 19	0. 00	35. 19
sloping 10 to 45 degrees	m2	0. 00	0. 00	40. 45	0. 00	40. 45
sloping 46 to 90 degrees	m2	0. 00	0. 00	45. 73	0. 00	45. 73
vertical	m2	0. 00	0. 00	45. 73	0. 00	45. 73

20mm two coat mastic asphalt to
BS1097, 225 to 300mm wide

flat	m2	0. 00	0. 00	25. 14	0. 00	25. 14
sloping 10 to 45 degrees	m2	0. 00	0. 00	28. 19	0. 00	28. 19
sloping 46 to 90 degrees	m2	0. 00	0. 00	32. 67	0. 00	32. 67
vertical	m2	0. 00	0. 00	32. 67	0. 00	32. 67

Damp-proofing and tanking (cont'd)	Unit	Labour hours	Net labour (£)	Net material (£)	O'heads /profit (£)	Total (£)
30mm three coat mastic asphalt to BS1097, over 300mm wide						
flat	m2	0.00	0.00	25.46	0.00	25.46
sloping 10 to 45 degrees	m2	0.00	0.00	29.26	0.00	29.26
sloping 46 to 90 degrees	m2	0.00	0.00	33.09	0.00	33.09
vertical	m2	0.00	0.00	33.09	0.00	33.09
30mm three coat mastic asphalt to BS1097, not exceeding 150mm wide						
flat	m2	0.00	0.00	31.17	0.00	31.17
sloping 10 to 45 degrees	m2	0.00	0.00	58.54	0.00	58.54
sloping 46 to 90 degrees	m2	0.00	0.00	66.18	0.00	66.18
vertical	m2	0.00	0.00	66.18	0.00	66.18
30mm three coat mastic to BS1097, 150 to 225mm wide						
flat	m2	0.00	0.00	44.54	0.00	44.54
sloping 10 to 45 degrees	m2	0.00	0.00	51.21	0.00	51.21
sloping 46 to 90 degrees	m2	0.00	0.00	68.64	0.00	68.64
vertical	m2	0.00	0.00	57.89	0.00	57.89
30mm three coat mastic asphalt to BS1097, 225 to 300mm wide						
flat	m2	0.00	0.00	31.81	0.00	31.81
sloping 10 to 45 degrees	m2	0.00	0.00	36.58	0.00	36.58
sloping 46 to 90 degrees	m2	0.00	0.00	41.35	0.00	41.35
vertical	m2	0.00	0.00	41.35	0.00	41.35
20mm two coat mastic asphalt to BS6577, over 300mm wide						
flat	m2	0.00	0.00	25.46	0.00	25.46
sloping 10 to 45 degrees	m2	0.00	0.00	29.26	0.00	29.26
sloping 46 to 90 degrees	m2	0.00	0.00	33.09	0.00	33.09
vertical	m2	0.00	0.00	32.09	0.00	32.09

ASPHALT TANKING

	Unit	Labour hours	Net labour (£)	Net material (£)	O'heads /profit (£)	Total (£)
20mm two coat mastic asphalt to BS6577, not exceeding 150mm wide						
flat	m2	0.00	0.00	51.17	0.00	51.17
sloping 10 to 45 degrees	m2	0.00	0.00	58.54	0.00	58.54
sloping 46 to 90 degrees	m2	0.00	0.00	61.18	0.00	61.18
vertical	m2	0.00	0.00	61.18	0.00	61.18
20mm two coat mastic asphalt to BS6577, 150 to 225mm wide						
flat	m2	0.00	0.00	44.54	0.00	44.54
sloping 10 to 45 degrees	m2	0.00	0.00	51.21	0.00	51.21
sloping 46 to 90 degrees	m2	0.00	0.00	68.64	0.00	68.64
vertical	m2	0.00	0.00	57.89	0.00	57.89
20mm two coat mastic asphalt to BS6577, 225 to 300mm wide						
flat	m2	0.00	0.00	31.81	0.00	31.81
sloping 10 to 45 degrees	m2	0.00	0.00	36.58	0.00	36.58
sloping 46 to 90 degrees	m2	0.00	0.00	41.35	0.00	41.35
vertical	m2	0.00	0.00	41.35	0.00	41.35

181

	Unit	Labour hours	Net labour (£)	Net material (£)	O'heads /profit (£)	Total (£)

J21 ASPHALT ROOFING

The following are specialist sub-contractor's prices and include 15% overheads and profit

Roofing

20mm two coat mastic asphalt to
BS988, over 300mm wide

flat	m2	0.00	0.00	22.90	0.00	22.90
sloping 10 to 45 degrees	m2	0.00	0.00	26.33	0.00	26.33
sloping 46 to 90 degrees	m2	0.00	0.00	29.78	0.00	29.78
vertical	m2	0.00	0.00	29.78	0.00	29.78

20mm two coat mastic asphalt to
BS988, not exceeding 150mm wide

flat	m2	0.00	0.00	45.67	0.00	45.67
sloping 10 to 45 degrees	m2	0.00	0.00	52.68	0.00	52.68
sloping 46 to 90 degrees	m2	0.00	0.00	59.56	0.00	59.56
vertical	m2	0.00	0.00	59.56	0.00	59.56

20mm two coat mastic asphalt to
BS988, 150 to 225mm wide

flat	m2	0.00	0.00	38.40	0.00	38.40
sloping 10 to 45 degrees	m2	0.00	0.00	43.90	0.00	43.90
sloping 46 to 90 degrees	m2	0.00	0.00	49.63	0.00	49.63
vertical	m2	0.00	0.00	49.63	0.00	49.63

20mm two coat mastic asphalt to
BS988, 225 to 300mm wide

flat	m2	0.00	0.00	27.26	0.00	27.26
sloping 10 to 45 degrees	m2	0.00	0.00	31.36	0.00	31.36
sloping 46 to 90 degrees	m2	0.00	0.00	35.44	0.00	35.44
vertical	m2	0.00	0.00	35.44	0.00	35.44

LIQUID DAMP-PROOF MEMBRANES

	Unit	Labour hours	Net labour (£)	Net material (£)	O'heads /profit (£)	Total (£)
J30 LIQUID DAMP-PROOF MEMBRANES						

Labour rate 8.00 per hour

Membranes

Two coats bituminous emulsion on concrete horizontal surfaces blinded with sand, width

| over 300mm | m2 | 0.35 | 2.80 | 1.90 | 0.70 | 5.40 |

Three coats bituminous emulsion on concrete vertical surface blinded with sand, width

| over 300mm | m2 | 0.40 | 3.20 | 2.86 | 0.91 | 6.97 |

RATES FOR MEASURED WORK

	Unit	Labour hours	Net labour (£)	Net material (£)	O'heads /profit (£)	Total (£)

J41 BUILT UP ROOFING

Labour rate 8.00 per hour

Built up bituminous felt roof coverings, layers fully bonded wit hot bitumen laid to 5 degrees pitch

Fibre based sand surfaced
felt type 1B weighing 14kg/10m2

one layer	m2	0. 22	1. 76	1. 45	0. 48	3. 69

Fibre based sand surfaced
felt type 1B weighing 18kg/10m2

one layer	m2	0. 23	1. 84	1. 72	0. 53	4. 09
two layers	m2	0. 30	2. 40	3. 71	0. 92	7. 03
three layers	m2	0. 45	3. 60	5. 73	1. 40	10. 73

Fibre based sand surfaced
felt type 1B weighing 25kg/10m2

one layer	m2	0. 24	1. 92	2. 28	0. 63	4. 83
two layers	m2	0. 32	2. 56	4. 84	1. 11	8. 51
three layers	m2	0. 47	3. 76	7. 51	1. 69	12. 96

Fibre based mineral surfaced
felt type 1E weighing 38kg/10m2

one layer	m2	0. 26	2. 08	3. 27	0. 80	6. 15

Glass fibre based sand surfaced
felt type 3B weighing 18kg/10m2

one layer	m2	0. 23	1. 84	1. 89	0. 56	4. 29
two layers	m2	0. 30	2. 40	4. 06	0. 97	7. 43
three layers	m2	0. 45	3. 60	6. 24	1. 48	11. 32

Glass fibre based mineral surfaced
felt type 3E weighing 28kg/10m2

one layer	m2	0. 26	2. 08	2. 60	0. 70	5. 38

BUILT UP ROOFING

	Unit	Labour hours	Net labour (£)	Net material (£)	O'heads /profit (£)	Total (£)
Glass fibre based venting layer felt type 3G weighing 28kg/10m2						
one layer	m2	0.28	2.24	3.00	0.79	6.03
Polyester based sand surfaced felt type 5U weighing 29kg/10m2						
one layer	m2	0.25	2.00	6.80	1.32	10.12
Polyester based sand surfaced felt type 5B weighing 34kg/10m2						
one layer	m2	0.26	2.08	5.69	1.17	8.94
Polyester based mineral surfaced felt type 5E weighing 38kg/10m2						
one layer	m2	0.28	2.24	6.70	1.34	10.28
Polyester based (180g/m2) sand surfaced elastomeric bitumen coated felt weighing 40kg/20m2						
one layer	m2	0.24	1.92	3.76	0.85	6.53
Polyester based (180g/m2) mineral surfaced elastomeric bitumen coated felt weighing 32kg/10m2						
one layer	m2	0.28	2.24	5.03	1.09	8.36
Euroroof Ltd, high performance, elastomeric fully bonded roofing system						
Two layer coverings, first layer HiTen G32, hot bitumen bonded to insulation layer, second layer cap sheet HiTen sanded finish P45	m2	0.40	3.20	8.51	1.76	13.47
top layer mineral surfaced	m2	0.10	0.80	1.03	0.27	2.10
10mm granite chippings	m2	0.10	0.80	3.88	0.70	5.38

Hiten G32 roofing (cont'd)	Unit	Labour hours	Net labour (£)	Net material (£)	O'heads /profit (£)	Total (£)
Skirtings; two layer; top layer mineral surfaces; dressed over tilting fillet; turned into grook						
not exceeding 200mm girth	m	0.25	2.00	1.87	0.58	4.45
200 to 400mm girth	m	0.35	2.80	3.74	0.98	7.52
Linings to gutters; three layers						
400 to 600mm girth average	m	0.65	5.20	7.46	1.90	14.56
Collars around large pipes; two layers						
150mm high	nr	0.35	2.80	0.53	0.50	3.83
Eurovent breather	nr	0.30	2.40	9.27	1.75	13.42
Euroroof pipe seal flashing for						
pipes 50mm to 75mm diameter	nr	0.30	2.40	14.54	2.54	19.48
pipes 100mm to 150mm diameter	nr	0.40	3.20	14.54	2.66	20.40

K

LININGS/SHEATHING/DRY PARTITIONING

MATERIAL COSTS

	Unit	Material supply (£)	Material waste (%)	Total (£)
Dry linings				
Gyproc wallboard				
9.5mm thick	m2	1.38	10.0	1.52
12.5mm thick	m2	1.59	10.0	1.75
Gyproc Fireline board				
12.5mm thick	m2	2.25	10.0	2.48
15mm thick	m2	2.75	10.0	3.03
Gyproc Duplex wallboard				
9.5mm thick	m2	1.94	10.0	2.13
12.5mm thick	m2	2.13	10.0	2.34
Gyproc Duplex Fireline				
12.5mm thick	m2	2.70	10.0	2.97
15mm thick	m2	3.28	10.0	3.61
Gyproc moisture resistant based				
9.5mm thick	m2	2.36	10.0	2.60
12.5mm thick	m2	2.78	10.0	3.06
Gyproc thermal board, standard				
25mm thick	m2	3.59	10.0	3.95
32mm thick	m2	4.13	10.0	4.54
Gyproc thermal board, vapour check				
40mm thick	m2	5.40	10.0	5.94
50mm thick	m2	6.20	10.0	6.82

RATES FOR MEASURED WORK

	Unit	Material supply (£)	Material waste (%)	Total (£)
Gyproc laminate				
25mm thick	m2	7. 35	10. 0	8. 08
32mm thick	m2	8. 30	10. 0	9. 13
Gyproc joint filler	bag	4. 20	10. 0	4. 62
Gyproc joint finish	bag	7. 25	10. 0	7. 98
Gyproc joint tape	roll	2. 31	10. 0	2. 54
Gyproc drywall angle bead	m	0. 84	10. 0	0. 92
Gyproc drywall edge bead				
9.5mm	m	0. 53	10. 0	0. 58
12.5mm	m	0. 55	10. 0	0. 61
15mm	m	0. 56	10. 0	0. 62
Flooring				
Softwood boarding butt jointed				
19 x 75mm	m2	6. 99	5. 0	7. 34
25 x 100mm	m2	7. 52	5. 0	7. 90
Softwood tongued and grooved boarding				
25 x 125mm	m2	7. 52	5. 0	7. 90
25 x 150mm	m2	7. 51	5. 0	7. 89
Chipboard flooring				
18mm	m2	3. 30	5. 0	3. 47
25mm	m2	4. 74	5. 0	4. 98
Partitions				
Gyproc drywall topcoat	m2	24. 20	10. 0	26. 62

MATERIAL COSTS

	Unit	Material supply (£)	Material waste (%)	Total (£)
Gyproc wallboard				
9.5mm thick	m2	1.38	10.0	1.52
12.5mm thick	m2	1.59	10.0	1.75
Gyproc Fireline board				
12.5mm thick	m2	2.25	10.0	2.48
15mm thick	m2	2.78	10.0	3.06
Gyproc Duplex wallboard				
9.5mm thick	m2	1.92	10.0	2.11
12.5mm thick	m2	2.19	10.0	2.41
Gyproc Duplex Fireline				
12.5mm thick	m2	2.70	10.0	2.97
15mm thick	m2	3.28	10.0	3.61
Gyproc moisture resistant based				
9.5mm thick	m2	2.36	10.0	2.60
12.5mm thick	m2	2.36	10.0	2.60
Paramount partition				
57mm thick	m2	5.30	10.0	5.83
63mm thick	m2	6.04	10.0	6.64
Gyproc thermal board, standard				
25mm thick	m2	3.59	10.0	3.95
32mm thick	m2	4.13	10.0	4.54
Gyproc thermal board, vapour check				
40mm thick	m2	5.41	10.0	5.95
50mm thick	m2	6.20	10.0	6.82

191

RATES FOR MEASURED WORK

	Unit	Material supply (£)	Material waste (%)	Total (£)
Gyproc laminate				
25mm thick	m2	7.35	10.0	8.08
32mm thick	m2	8.30	10.0	9.13
Gyproc joint filler	bag	4.20	10.0	4.62
Gyproc joint finish	bag	7.14	10.0	7.85
Gyproc joint tape	roll	2.31	10.0	2.54
Gyproc drywall angle bead	m	0.84	10.0	0.92
Gyproc drywall edge bead				
9.5mm	m	0.53	10.0	0.58
12.5mm	m	0.55	10.0	0.61
15mm	m	0.56	10.0	0.62
Flooring				
Softwood boarding butt jointed				
19 x 75mm	m2	6.99	5.0	7.34
25 x 100mm	m2	7.52	5.0	7.90
Softwood tongued and grooved boarding				
25 x 125mm	m2	7.52	5.0	7.90
25 x 150mm	m2	7.51	5.0	7.89
Chipboard flooring				
18mm	m2	3.30	5.0	3.47
25mm	m2	4.74	5.0	4.98
Partitions				
Gyproc drywall topcoat	m2	24.20	10.0	26.62

MATERIAL COSTS

	Unit	Material supply (£)	Material waste (%)	Total (£)
Wallboard				
9.5mm thick	m2	1.38	10.0	1.52
12.5mm thick	m2	1.59	10.0	1.75
Duplex wallboard				
9.5mm thick	m2	1.92	10.0	2.11
12.5mm thick	m2	2.17	10.0	2.39
Fireline board				
12.5mm thick	m2	2.70	10.0	2.97
15mm thick	m2	3.28	10.0	3.61
Ceiling tiles				
Armstrong ceiling tiles				
Synonymes				
Omega, white	m2	12.13	5.0	12.74
Sigma, white	m2	11.65	5.0	12.23
Gamma, white	m2	13.24	5.0	13.90
Gammatone, coloured	m2	17.51	5.0	18.39
Delta	m2	10.65	5.0	11.18
Visual				
V49, colour	m2	24.54	5.0	25.77
V49, white	m2	22.44	5.0	23.56
H36, colour	m2	24.38	5.0	25.60
H36, white	m2	22.44	5.0	23.56
V64, colour	m2	24.60	5.0	25.83
V64, white	m2	23.30	5.0	24.47
Microlook				
Finesse, white	m2	9.24	5.0	9.70
Finesse, coloured	m2	10.91	5.0	11.46
Dune, white	m2	7.43	5.0	7.80
Dune, coloured	m2	7.86	5.0	8.25
Galaxie, white	m2	14.07	5.0	14.77
Galaxie, coloured	m2	14.82	5.0	15.56
Doric, white	m2	13.50	5.0	14.18
Doric, coloured	m2	14.26	5.0	14.97
Decade, white	m2	13.50	5.0	14.18

RATES FOR MEASURED WORK

	Unit	Material supply (£)	Material waste (%)	Total (£)
Decade, coloured	m2	14.26	5.0	14.97
Finesse Step, white	m2	12.00	5.0	12.60
Cirrus Step, white	m2	10.29	5.0	10.80
Galaxie Step, white	m2	13.94	5.0	14.64
Galaxie Step, coloured	m2	14.71	5.0	15.45
Decade Parallel, white	m2	13.90	5.0	14.60
Decade Parallel, coloured	m2	14.75	5.0	15.49
Doric Parallel, white	m2	14.78	5.0	15.52
Doric Parallel, coloured	m2	10.36	5.0	10.88
Citrus Parallel, white	m2	14.78	5.0	15.52
Finesse Parallel, white	m2	10.36	5.0	10.88
Galaxie Parallel, white	m2	14.77	5.0	15.51
Fissured Natural Parallel, white	m2	12.40	5.0	13.02
Cortega Sektor, white	m2	7.68	5.0	8.06
Cortega Segment, white	m2	7.79	5.0	8.18
Cortega Labyrinth, white	m2	7.68	5.0	8.06
Radius 1M, white	m2	10.90	5.0	11.45
Radius 1M, coloured	m2	11.51	5.0	12.09
Radius GL, white	m2	10.31	5.0	10.83
Radius GL, coloured	m2	11.55	5.0	12.13
Image, white	m2	12.50	5.0	13.13
Gleis, white	m2	12.13	5.0	12.74
Munster, white	m2	13.86	5.0	14.55
Munster, coloured	m2	14.60	5.0	15.33

Armstrong Supralook, horizontal

	Unit	Material supply (£)	Material waste (%)	Total (£)
Finesse I, white	m2	13.57	5.0	14.25
Finesse III, white	m2	13.57	5.0	14.25
Finesse IV, white	m2	13.31	5.0	13.98

Secondlook

	Unit	Material supply (£)	Material waste (%)	Total (£)
Cortega I, white	m2	6.75	5.0	7.09
Cortega I, coloured	m2	7.49	5.0	7.86
Cortega III & IV, white	m2	7.14	5.0	7.50
Cortega III & IV, coloured	m2	7.58	5.0	7.96
Galaxie I & III, white	m2	14.35	5.0	15.07
Galaxie IV, white	m2	14.26	5.0	14.97
Samserra I, white	m2	14.67	5.0	15.40
Samserra III, white	m2	14.70	5.0	15.43
Samserra IV, white	m2	14.84	5.0	15.58

PLASTERBOARD DRY LINING

	Unit	Labour hours	Net labour (£)	Net material (£)	O'heads /profit (£)	Total (£)

K10 PLASTERBOARD DRY LINING

Labour rate 10.33 per hour

Baseboard, Gyproc wallboard square edge, 1200 x 2400mm, taped butt joints, to receive skim coat, to timber with nails

9.5mm thick

Walls

height 2.10 to 2.40m	m	0.53	5.47	4.11	1.44	11.02
height 2.40 to 2.70m	m	0.59	6.09	4.62	1.61	12.32
height 2.70 to 3.00m	m	0.66	6.82	5.09	1.79	13.70
height 3.00 to 3.30m	m	0.73	7.54	5.65	1.98	15.17
height 3.30 to 3.60m	m	0.79	8.16	6.16	2.15	16.47
Ceilings, generally	m2	0.26	2.69	1.71	0.66	5.06

Sides, soffits or tops of isolated beams

girth not exceeding 600mm	m	0.16	1.65	1.03	0.40	3.08
girth 600 to 1200mm	m	0.31	3.20	2.06	0.79	6.05
girth 1200 to 1800mm	m	0.47	4.86	3.08	1.19	9.13

Sides of isolated columns

girth not exceeding 600mm	m	0.13	1.34	1.03	0.36	2.73
girth 600 to 1200mm	m	0.26	2.69	2.06	0.71	5.46
girth 1200 to 1800mm	m	0.40	4.13	3.08	1.08	8.29

Add to the previous prices for Duplex wallboard	m2	0.00	0.00	0.21	0.03	0.24
Add to the previous prices for moisture resistant board	m2	0.00	0.00	0.39	0.06	0.45

Increase the labour element of ceilings and beams for work over 3.5m high by 10%

RATES FOR MEASURED WORK

	Unit	Labour hours	Net labour (£)	Net material (£)	O'heads /profit (£)	Total (£)
12.5mm thick						
Walls						
height 2.10 to 2.40m	m	0.62	6.40	4.71	1.67	12.78
height 2.40 to 2.70m	m	0.70	7.23	5.25	1.87	14.35
height 2.70 to 3.00m	m	0.78	8.06	5.81	2.08	15.95
height 3.00 to 3.30m	m	0.86	8.88	6.02	2.23	17.13
height 3.30 to 3.60m	m	0.94	9.71	6.99	2.50	19.20
Ceilings, generally	m2	0.31	3.20	1.94	0.77	5.91
Sides, soffits or tops of isolated beams						
girth not exceeding 600mm	m	0.19	1.96	1.17	0.47	3.60
girth 600 to 1200mm	m	0.37	3.82	2.31	0.92	7.05
girth 1200 to 1800mm	m	0.56	5.78	3.50	1.39	10.67
Sides of isolated columns						
girth not exceeding 600mm	m	0.16	1.65	1.17	0.42	3.24
girth 600 to 1200mm	m	0.32	3.31	2.33	0.85	6.49
girth 1200 to 1800mm	m	0.48	4.96	3.50	1.27	9.73
Add to previous prices for Duplex wallboard	m2	0.00	0.00	0.21	0.03	0.24
Add to previous prices for moisture resistant board	m2	0.00	0.00	0.47	0.01	0.48

Increase the labour element of all
items for working in staircase
areas or compartments not exceeding 4m2 by 15%

Increase the labour element of
ceilings and beams for work over
3.5m high by 10%

PLASTERBOARD DRY LINING

	Unit	Labour hours	Net labour (£)	Net material (£)	O'heads /profit (£)	Total (£)
Baseboarding, Gyproc thermal board tapered edge, 1200 x 2400mm, taped butt joints, to receive skmm coat, to timber with nails						
25mm thick						
Walls						
height 2.10 to 2.40m	m	0.75	7.75	9.95	2.65	20.35
height 2.40 to 2.70m	m	0.84	8.68	11.19	2.98	22.85
height 2.70 to 3.00m	m	0.93	9.61	12.44	3.31	25.36
height 3.00 to 3.30m	m	1.02	10.54	13.69	3.63	27.86
height 3.30 to 3.60m	m	1.12	11.57	14.93	3.97	30.47
Ceilings, generally	m2	0.36	3.72	4.15	1.18	9.05
Sides, soffits or tops of isolated beams						
girth not exceeding 600mm	m	0.22	2.27	2.49	0.71	5.47
girth 600 to 1200mm	m	0.44	4.55	4.98	1.43	10.96
girth 1200 to 1800mm	m	0.66	6.82	7.67	2.17	16.66
Sides of isolated columns						
girth not exceeding 600mm	m	0.19	1.96	2.48	0.67	5.11
girth 600 to 1200mm	m	0.38	3.93	4.98	1.34	10.25
girth 1200 to 1800mm	m	0.57	5.89	7.47	2.00	15.36
Add to the previous prices for tapered vapour check thermal board	m2	0.00	0.00	0.24	0.04	0.28
Add to the previous prices for tapered urethane laminate board	m2	0.00	0.00	1.44	0.22	1.66

Increase the labour element of all
items for working in staircase
areas or compartments not
exceeding 4m2 by 15%

RATES FOR MEASURED WORK

	Unit	Labour hours	Net labour (£)	Net material (£)	O'heads /profit (£)	Total (£)
Increase the labour elements of ceilings and beams for work over 3.5m high by 10%						
40mm thick						
Walls						
height 2.10 to 2.40m	m	0.86	8.88	14.74	3.54	27.16
height 2.40 to 2.70m	m	0.97	10.02	16.59	3.99	30.60
height 2.70 to 3.00m	m	1.08	11.16	18.43	4.44	34.03
height 3.00 to 3.30m	m	1.19	12.29	20.29	4.89	37.47
height 3.30 to 3.60m	m	1.30	13.43	22.11	5.33	40.87
Ceilings, generally	m2	0.41	4.24	6.14	1.56	11.94
Sides, soffits or tops of isolated beams						
girth not exceeding 600mm	m	0.25	2.58	3.69	0.94	7.21
girth 600 to 1200mm	m	0.50	5.17	7.37	1.88	14.42
girth 1200 to 1800mm	m	0.75	7.75	11.06	2.82	21.63
Sides of isolated columns						
girth not exceeding 600mm	m	0.22	2.27	3.69	0.89	6.85
girth 600 to 1200mm	m	0.44	4.55	7.37	1.79	13.71
girth 1200 to 1800mm	m	0.66	6.82	11.06	2.68	20.56

Increase the labour element of all items for working in staircase areas or compartments not exceeding 4m2 by 15%

Increase the labour element of ceilings and beams for work over 3.5m high by 10%

TIMBER BOARD FLOORINGS

	Unit	Labour hours	Net labour (£)	Net material (£)	O'heads /profit (£)	Total (£)
K20 TIMBER BOARD FLOORINGS						
Labour rate 8.00 per hour						
Butt jointed boarding to joists, size						
25 x 125mm	m2	0.95	7.60	6.34	2.09	16.03
25 x 100mm	m2	0.82	6.56	7.90	2.17	16.63
Tongued and grooved boarding to joists, size						
25 x 125mm	m	1.16	9.28	7.90	2.58	19.76
25 x 150mm	m	1.10	8.80	7.89	2.50	19.19
Chipboard boarding to floors and roofs, thickness						
18mm	m	0.44	3.52	4.66	1.23	9.41
25mm	m	0.50	4.00	4.97	1.35	10.32

	Unit	Labour hours	Net labour (£)	Net material (£)	O'heads /profit (£)	Total (£)

K31 PARTITIONS/LININGS

Labour rate 10.33 per hour

50mm thick, 19mm Gyproc square edge plank core, 12.5mm Gyproc tapered edge wallboard facing both sides bonded with adhesive, joints taped and filled, one coat Gyproc drywall topcoat

Walls

height 2.10 to 2.40m	m	2.10	21.69	19.42	6.17	47.28
height 2.40 to 2.70m	m	2.40	24.79	21.86	7.00	53.65
height 2.70 to 3.00m	m	2.70	27.89	24.29	7.83	60.01
height 3.00 to 3.30m	m	3.00	30.99	26.71	8.65	66.35
height 3.30 to 3.60m	m	3.30	34.09	29.15	9.49	72.73

65mm thick, 19mm Gyproc square edge plank core, 19mm Gyproc tapered edge plank facing both sides bonded with adhesive, joints taped and filled, one coat Gyproc drywall topcoat

Walls

height 2.10 to 2.40m	m	2.52	26.03	25.20	7.68	58.91
height 2.40 to 2.70m	m	2.83	29.23	28.30	8.63	66.16
height 2.70 to 3.00m	m	3.15	32.54	31.50	9.61	73.65
height 3.00 to 3.30m	m	3.47	35.85	34.65	10.57	81.07
height 3.30 to 3.60m	m	3.78	39.05	37.80	11.53	88.38

PARTITIONS/LININGS

	Unit	Labour hours	Net labour (£)	Net material (£)	O'heads /profit (£)	Total (£)

Partitions, Paramount dry partition, Gyproc wallboard on cellular core 57mm thick, 38 x 38mm battens, grey faced with square butt joints for plaster (not included)

Walls

height 2.10 to 2.40m	m	2.04	21.07	15.82	5.53	42.42
height 2.40 to 2.70m	m	2.30	23.76	17.81	6.24	47.81
height 2.70 to 3.00m	m	2.55	26.34	19.78	6.92	53.04
height 3.00 to 3.30m	m	2.80	28.92	21.76	7.60	58.28
height 3.30 to 3.60m	m	3.06	31.61	23.74	8.30	63.65

Partitions, Paramount dry partition, Gyproc wallboard on cellular core, 63mm thick, 38 x 38mm jointing battens, grey faced with square butt joints for plaster (not included)

Walls

height 2.10 to 2.40m	m	2.16	22.31	18.22	6.08	46.61
height 2.40 to 2.70m	m	2.43	25.10	20.21	6.80	52.11
height 2.70 to 3.00m	m	2.70	27.89	22.77	7.60	58.26
height 3.00 to 3.30m	m	2.97	30.68	25.16	8.38	64.22
height 3.30 to 3.60m	m	3.24	33.47	27.32	9.12	69.91

Perimeter fixing batten

30 x 19mm

to timber with nails	m	0.12	1.24	0.24	0.22	1.70
to brickwork or concrete with screws, plugging	m	0.18	1.86	0.35	0.33	2.54

38 x 19mm

to timber with nails	m	0.13	1.34	0.28	0.24	1.86
to brickwork or concrete with screws, plugging	m	0.19	1.96	0.39	0.35	2.70

Partitions (cont'd)	Unit	Labour hours	Net labour (£)	Net material (£)	O'heads /profit (£)	Total (£)

Right angle junction

50mm thick, 30 x 38mm and 30 x 19mm battens	m	0.40	4.13	0.78	0.74	5.65
57mm thick, 38 x 38mm and 38 x 19mm battens	m	0.43	4.44	0.83	0.79	6.06
63mm thick, 38 x 38mm and 38 x 19mm battens	m	0.44	4.55	0.83	0.81	6.19

T junction

50mm thick, 30 x 38mm and 30 x 19mm battens	m	0.33	3.41	0.79	0.63	4.83
57mm thick, 38 x 38mm and 38 x 19mm battens	m	0.34	3.51	0.83	0.65	4.99
63mm thick, 38 x 38mm and 38 x 19mm battens	m	0.35	3.62	0.83	0.67	5.12

DRY LININGS

Thistlebond system of dry lining, Gyproc tapered edge wallboard on gypsum adhesive dabs with alignment pads, joints taped and filled, one coat Gyproc drywall topcoat

Board size 900 x 2400 x 9.5mm thick to walls

height 2.10 to 2.40m	m	0.79	8.16	5.88	2.11	16.15
height 2.40 to 2.70m	m	0.89	9.19	6.60	2.37	18.16
height 2.70 to 3.00m	m	0.99	10.23	7.34	2.64	20.21
height 3.00 to 3.30m	m	1.09	11.26	8.07	2.90	22.23
height 3.30 to 3.60m	m	1.19	12.29	8.81	3.17	24.27

Board size 900 x 2400 x 12.5mm thick to walls

height 2.10 to 2.40m	m	0.99	10.23	6.43	2.50	19.16
height 2.40 to 2.70m	m	1.05	10.85	7.23	2.71	20.79
height 2.70 to 3.00m	m	1.17	12.09	8.03	3.02	23.14
height 3.00 to 3.30m	m	1.29	13.33	8.84	3.33	25.50
height 3.30 to 3.60m	m	1.40	14.46	9.64	3.62	27.72

PARTITIONS/LININGS

	Unit	Labour hours	Net labour (£)	Net material (£)	O'heads /profit (£)	Total (£)

Thistlebond TL system of dry lining, Gyproc tapered edge plasterboard on Gypsum adhesive dabs with alignment pads, Gyproc nailable plug secondary fixing, joints taped and filled, one coat Gyproc drywall topcoat, Gyproc thermal board, size 1200 x 2400m

25mm thick to walls

	Unit	Labour hours	Net labour	Net material	O'heads /profit	Total
height 2.10 to 2.40m	m	0.84	8.68	7.92	2.49	19.09
height 2.40 to 2.70m	m	0.95	9.81	10.00	2.97	22.78
height 2.70 to 3.00m	m	1.05	10.85	11.12	3.30	25.27
height 3.00 to 3.30m	m	1.16	11.98	10.88	3.43	26.29
height 3.30 to 3.60m	m	1.26	13.02	13.34	3.95	30.31

32mm thick to walls

	Unit	Labour hours	Net labour	Net material	O'heads /profit	Total
height 2.10 to 2.40m	m	0.86	8.88	8.25	2.57	19.70
height 2.40 to 2.70m	m	0.97	10.02	9.84	2.98	22.84
height 2.70 to 3.00m	m	1.08	11.16	10.93	3.31	25.40
height 3.00 to 3.30m	m	1.19	12.29	12.02	3.65	27.96
height 3.30 to 3.60m	m	1.30	13.43	13.33	4.01	30.77

40mm thick to walls

	Unit	Labour hours	Net labour	Net material	O'heads /profit	Total
height 2.10 to 2.40m	m	0.89	9.19	9.58	2.82	21.59
height 2.40 to 2.70m	m	1.00	10.33	10.77	3.17	24.27
height 2.70 to 3.00m	m	1.10	11.36	11.97	3.50	26.83
height 3.00 to 3.30m	m	1.22	12.60	13.17	3.87	29.64
height 3.30 to 3.60m	m	1.33	13.74	14.36	4.21	32.31

50mm thick to walls

	Unit	Labour hours	Net labour	Net material	O'heads /profit	Total
height 2.10 to 2.40m	m	0.91	9.40	10.31	2.96	22.67
height 2.40 to 2.70m	m	1.03	10.64	11.59	3.33	25.56
height 2.70 to 3.00m	m	1.14	11.78	12.88	3.70	28.36
height 3.00 to 3.30m	m	1.25	12.91	14.18	4.06	31.15
height 3.30 to 3.60m	m	1.37	14.15	15.46	4.44	34.05

RATES FOR MEASURED WORK

	Unit	Labour hours	Net labour (£)	Net material (£)	O'heads /profit (£)	Total (£)
Gyproc urethane laminate board, size 1200 x 2400mm						
25mm thick to walls						
height 2.10 to 2.40m	m	0.84	8.68	13.40	3.31	25.39
height 2.40 to 2.70m	m	0.95	9.81	15.08	3.73	28.62
height 2.70 to 3.00m	m	1.05	10.85	16.76	4.14	31.75
height 3.00 to 3.30m	m	1.16	11.98	18.44	4.56	34.98
height 3.30 to 3.60m	m	1.26	13.02	20.11	4.97	38.10
32mm thick to walls						
height 2.10 to 2.40m	m	0.86	8.88	17.44	3.95	30.27
height 2.40 to 2.70m	m	0.97	10.02	19.61	4.44	34.07
height 2.70 to 3.00m	m	1.08	11.16	21.80	4.94	37.90
height 3.00 to 3.30m	m	1.19	12.29	23.98	5.44	41.71
height 3.30 to 3.60m	m	1.30	13.43	26.16	5.94	45.53
40mm thick to walls						
height 2.10 to 2.40m	m	0.59	6.09	18.92	3.75	28.76
height 2.40 to 2.70m	m	1.00	10.33	21.29	4.74	36.36
height 2.70 to 3.00m	m	1.10	11.36	23.66	5.25	40.27
height 3.00 to 3.30m	m	1.22	12.60	26.02	5.79	44.41
height 3.30 to 3.60m	m	1.33	13.74	28.39	6.32	48.45

SUSPENDED CEILINGS

	Unit	Labour hours	Net labour (£)	Net material (£)	O'heads /profit (£)	Total (£)

K40 SUSPENDED CEILINGS

Labour rate 10.33 per hour

Gypsum Paraclip system

British Gypsum Paraclip metal grid
system, comprising P502 primary
channel sections, P506 main
supporting sections, P507N cross
supporting sections, suspension
height 500mm, with clips and
brackets, nailed to timber soffits
or wired to steelwork

Horizontal, board 9.5mm thick
Gyproc Industrial grade wallboards

600 x 1800mm	m2	0. 40	4. 13	6. 60	1. 61	12. 34
600 x 2400mm	m2	0. 38	3. 93	6. 60	1. 58	12. 11
600 x 2700mm	m2	0. 36	3. 72	6. 60	1. 55	11. 87
600 x 3000mm	m2	0. 34	3. 51	6. 60	1. 52	11. 63

Horizontal board 12.5mm thick
Gyproc Industrial grade wallboard

600 x 1800mm	m2	0. 42	4. 34	6. 92	1. 69	12. 95
600 x 2400mm	m2	0. 40	4. 13	6. 92	1. 66	12. 71
600 x 2700mm	m2	0. 38	3. 93	6. 92	1. 63	12. 48
600 x 3000mm	m2	0. 36	3. 72	6. 92	1. 60	12. 24

Add to the previous prices for
plugging and screwing to
concrete soffits | m2 | 0. 08 | 0. 83 | 0. 17 | 0. 03 | 1. 03

Increase the labour element for
working in staircase areas or
compartments not exceeding 4m2
by 15%

Increase the labour element for
work over 3.5m high by 10%

RATES FOR MEASURED WORK

	Unit	Labour hours	Net labour (£)	Net material (£)	O'heads /profit (£)	Total (£)
British Gypsum Gyproc M/F system, comprising ref MF7 primary support channels, ref M/F ceiling sections ref MF8 strap hangers, suspension height 500mm, screwed to timber soffits						
Single layer of 900 x 1800mm tapered edge panels, joints taped and filled						
Horizontal						
wallboard, 12.5mm thick	m2	0.72	7.44	5.33	1.92	14.69
Duplex wallboard, 12.5mm thick	m2	0.72	7.44	5.38	1.92	14.74
Fireline board, 12.5mm thick	m2	0.72	7.44	6.27	2.06	15.77
Vertical						
wallboard, 12.5mm thick	m2	0.74	7.64	5.33	1.95	14.92
Duplex wallboard, 12.5mm thick	m2	0.74	7.64	5.38	1.95	14.97
Fireline board, 12.5mm thick	m2	0.74	7.64	6.27	2.09	16.00
Double layer of 900 x 1800mm panels, first layer butt jointed, second layer tapered edge panels, joints taped and filled						
Horizontal						
wallboard, 12.5mm thick	m2	1.01	10.43	7.28	2.66	20.37
Duplex wallboard, 12.5mm thick	m2	1.01	10.43	7.70	2.72	20.85
Fireline board, 12.5mm thick	m2	1.01	10.43	9.15	2.94	22.52
Vertical						
wallboard, 12.5mm thick	m2	1.03	10.64	7.28	2.69	20.61
Duplex wallboard, 12.5mm thick	m2	1.03	10.64	7.70	2.75	21.09
Fireline board, 12.5mm thick	m2	1.03	10.64	9.15	2.97	22.76
Add to the previous prices for plugging and screwing to concrete soffits	m2	0.08	0.83	0.16	0.02	1.01

SUSPENDED CEILINGS

	Unit	Labour hours	Net labour (£)	Net material (£)	O'heads /profit (£)	Total (£)
Add to the previous prices for drilling and screwing to steel-work	m2	0.15	1.55	0.38	0.05	1.98

Increase the labour element for working in staircase areas or compartments not exceeding 4m2 by 15%

Increase the labour element for work over 3.5m high by 10%

Armstrong suspended ceilings, Trulok Microlock 15 suspension system with main runners and cross tees to form a 600 x 600mm module, suspension height 500mm, with all clips and brackets, nailed to timber soffits or wired to steel-work, quantities 250 to 1250m2

Horizontal

	Unit	Labour hours	Net labour	Net material	O'heads /profit	Total
Dune, white	m2	0.44	4.55	10.98	2.33	17.86
Dune, coloured	m2	0.44	4.55	11.45	2.40	18.40
Labyrinth, white	m2	0.44	4.55	11.26	2.37	18.18
Cortega, white	m2	0.44	4.55	10.15	2.21	16.91
Segment, white	m2	0.44	4.55	11.37	2.39	18.31
Sektor, white	m2	0.44	4.55	11.26	2.37	18.18

Vertical

	Unit	Labour hours	Net labour	Net material	O'heads /profit	Total
Dune, white	m2	0.48	4.96	10.98	2.39	18.33
Dune, coloured	m2	0.48	4.96	11.45	2.46	18.87
Labyrinth, white	m2	0.48	4.96	11.26	2.43	18.65
Cortega, white	m2	0.44	4.55	10.15	2.21	16.91
Segment, white	m2	0.48	4.96	11.37	2.45	18.78
Sektor, white	m2	0.48	4.96	11.26	2.43	18.65

L

WINDOWS/DOORS/STAIRS

ESTIMATES NOT GUESTIMATES

In the construction industry, when 20% of the design is fixed, so is 80% of the cost.

This means effective cost control is vital from the outset of every plumbing, electrical or minor works project.

As experienced quantity surveyors and cost engineers, Tweeds can achieve the optimum balance between scarce resources and desired benefits.

These Value Manage-ment skills turn outlines or concepts into a comprehensive vision which will stand the test of time.

Backed by extensive technical knowledge and the latest cost modelling techniques, we can set a framework to control financial discipline during the entire project.

So if you are seeking sound information on which to base your decisions, Tweeds provides forecasts you can rely on absolutely.

Tweeds

CHARTERED QUANTITY SURVEYORS,

LONDON   LIVERPOOL   MANCHESTER   KENDAL

NEWCASTLE-UPON-TYNE   MOLD   DUDLEY   PRESTON

RATES FOR MEASURED WORK

	Unit	Material supply (£)	Material waste (%)	Total (£)

Metal windows

Crittalls Duralife Homelight
standard galvanized steel windows,
type

NC5	nr	12.42	5.0	13.04
ND5	nr	14.18	5.0	14.89
NE13E	nr	39.13	5.0	41.09
NC13	nr	19.19	5.0	20.15
NE3	nr	49.07	5.0	51.52
NC13F	nr	44.14	5.0	46.35
ND13	nr	22.25	5.0	23.36
NC01	nr	34.91	5.0	36.66
NC5F	nr	28.81	5.0	30.25
ND5F	nr	30.62	5.0	32.15
NC05F	nr	29.68	5.0	31.16
ND1	nr	37.54	5.0	39.42
ND4F	nr	97.03	5.0	101.88
NDV5F	nr	33.81	5.0	35.50

Doors

Standard flush door plywood faced
both sides

35mm thick, size				
686 x 1981mm	nr	19.04	5.0	19.99
762 x 1981mm	nr	19.30	5.0	20.27
40mm thick, size				
626 x 2040mm	nr	20.05	5.0	21.05
726 x 2040mm	nr	20.28	5.0	21.29
826 x 2040mm	nr	21.18	5.0	22.24

Standard flush door sapele faced
both sides

35mm, thick size				
686 x 1981mm	nr	21.66	5.0	22.74
762 x 1981mm	nr	21.89	5.0	22.98

211

	Unit	Material supply (£)	Material waste (%)	Total (£)
40mm thick, size				
626 x 2040mm	nr	22. 39	5. 0	23. 51
726 x 2040mm	nr	22. 69	5. 0	23. 82
826 x 2040mm	nr	23. 55	5. 0	24. 73
Standard flush door teak faced both sides				
35mm thick, size				
686 x 1981mm	nr	38. 61	5. 0	40. 54
762 x 1981mm	nr	39. 83	5. 0	41. 82
40mm thick, size				
626 x 2040mm	nr	42. 67	5. 0	44. 80
726 x 2040mm	nr	42. 81	5. 0	44. 95
826 x 2040mm	nr	43. 76	5. 0	45. 95
Standard flush door, half hour fire check, sapele faced both sides, 44mm thick, size				
610 x 1981mm	nr	44. 72	5. 0	46. 96
686 x 1981mm	nr	44. 72	5. 0	46. 96
762 x 1981mm	nr	46. 48	5. 0	48. 80
838 x 2040mm	nr	46. 48	5. 0	48. 80
Standard flush door, half hour fire check, plywood faced both sides, 44mm thick, size				
686 x 1981mm	nr	32. 04	5. 0	33. 64
762 x 1981mm	nr	32. 04	5. 0	33. 64
762 x 2040mm	nr	33. 62	5. 0	35. 30
838 x 2040mm	nr	35. 07	5. 0	36. 82
Framed ledged and braced door 44mm thick with 19mm matchboarding				
686 x 1981mm	nr	52. 14	5. 0	54. 75
762 x 1981mm	nr	53. 14	5. 0	55. 80

MATERIAL COSTS

	Unit	Material supply (£)	Material waste (%)	Total (£)
Framed ledged and braced door 44mm thick with 25mm matchboarding				
813 x 2032mm	nr	53.14	5.0	55.80
838 x 1981mm	nr	53.14	5.0	55.80
Door lining size 32 x 142mm with loose stops for door, size				
686 x 1981mm	nr	14.64	5.0	15.37
762 x 1981mm	nr	14.64	5.0	15.37
838 x 1981mm	nr	14.87	5.0	15.61
Door lining size 32 x 140mm with loose stops for door, size				
686 x 1981mm	nr	17.49	5.0	18.36
762 x 1981mm	nr	17.49	5.0	18.36
838 x 1981mm	nr	17.49	5.0	18.36
Door frame, rebated, size 48 x 114mm for door size 762 x 1981mm	nr	23.50	5.0	24.68
Door frame, rebated, size 38 x 140mm for door size 762 x 1981mm	nr	27.85	5.0	29.24

Softwood windows

Standard softwood windows, with 75 x 150mm sills, ironmongery, without glazing bars, type

	Unit	Material supply (£)	Material waste (%)	Total (£)
N07V, 488 x 750mm	nr	33.53	5.0	35.21
N09V, 588 x 900mm	nr	34.93	5.0	36.68
N12V, 488 x 1200mm	nr	37.75	5.0	39.64
107C, 630 x 750mm	nr	38.77	5.0	40.71
110C, 630 x 1050mm	nr	42.83	5.0	44.97
112C, 630 x 1200mm	nr	44.88	5.0	47.12
110T, 630 x 1050mm	nr	59.14	5.0	62.10
113T, 630 x 1350mm	nr	63.18	5.0	66.34
109V, 630 x 900mm	nr	38.75	5.0	40.69
112V, 630 x 1200mm	nr	41.25	5.0	43.31
212DG, 1200 x 1200mm	nr	36.59	5.0	38.42
212W, 1200 x 1200mm	nr	60.63	5.0	63.66

RATES FOR MEASURED RATES

	Unit	Material supply (£)	Material waste (%)	Total (£)
210C, 1200 x 1050mm	nr	57. 18	5. 0	60. 04
212T, 1200 x 1200mm	nr	73. 51	5. 0	77. 19
212CV, 1200 x 1200mm	nr	76. 29	5. 0	80. 10
310CVC, 1770 x 1050mm	nr	106. 49	5. 0	111. 81
413CWC, 2339 x 1350mm	nr	138. 99	5. 0	145. 94

Hardwood windows

Standard hardwood windows without
glazing bars, type

	Unit	Material supply (£)	Material waste (%)	Total (£)
H2N10W, 915 x 1050mm	nr	100. 94	5. 0	105. 99
H2N13W, 915 x 1350mm	nr	105. 69	5. 0	110. 97
H2N13W, 915 x 1500mm	nr	105. 69	5. 0	110. 97
H210DG, 1200 x 1050mm	nr	73. 53	5. 0	77. 21
H213W, 1200 x 1350mm	nr	123. 18	5. 0	129. 34
H215W, 1200 x 1500mm	nr	126. 14	5. 0	132. 45
H2O9CV, 1200 x 900mm	nr	152. 51	5. 0	160. 14
H213CV, 1200 x 1300mm	nr	166. 73	5. 0	175. 07
H309CC, 1770 x 900mm	nr	189. 23	5. 0	198. 69
H312CC, 1770 x 1200mm	nr	206. 39	5. 0	216. 71
H307C, 1770 x 750mm	nr	136. 00	5. 0	142. 80
H312C, 1770 x 1200mm	nr	147. 11	5. 0	154. 47

Timber stairs

Closed tread staircase in one
flight, 2600mm height, consisting
of 12 treads and 13 risers, width

	Unit	Material supply (£)	Material waste (%)	Total (£)
864mm	nr	250. 17	5. 0	262. 68
914mm	nr	264. 88	5. 0	278. 12

Open tread staircase in one flight
2600mm high, consisting of 12
treads and 13 risers, width

	Unit	Material supply (£)	Material waste (%)	Total (£)
864mm	nr	452. 93	5. 0	475. 58
914mm	nr	470. 19	5. 0	493. 70

MATERIAL COSTS

	Unit	Material supply (£)	Material waste (%)	Total (£)
Glazing				
Clear float glass, thickness				
3mm	m2	25. 56	5. 0	26. 84
4mm	m2	28. 97	5. 0	30. 42
5mm	m2	42. 05	5. 0	44. 15
6mm	m2	47. 00	5. 0	49. 35
10mm	m2	99. 88	5. 0	104. 87
White patterned glass				
4mm	m2	21. 43	5. 0	22. 50
5mm	m2	30. 08	5. 0	31. 58
6mm	m2	38. 38	5. 0	40. 30
Rough cast glass				
6mm	m2	33. 56	5. 0	35. 24
10mm	m2	48. 53	5. 0	50. 96
Georgian wired cast glass				
7mm	m2	35. 44	5. 0	37. 21
Georgian wired polished glass				
6mm	m2	77. 27	5. 0	81. 13

RATES FOR MEASURED WORK

	Unit	Labour hours	Net labour (£)	Net material (£)	O'heads /profit (£)	Total (£)
L10 TIMBER WINDOWS						
Labour rate 8.00 per hour						
Standard softwood windows without glazing bars, type						
N07V, 488 x 750mm	nr	0.75	6.00	35.21	6.18	47.39
N09V, 488 x 900mm	nr	1.00	8.00	36.67	6.70	51.37
N12V, 488 x 1200mm	nr	1.25	10.00	39.64	7.45	57.09
107C, 630 x 750mm	nr	0.75	6.00	40.71	7.01	53.72
110C, 630 x 1050mm	nr	1.00	8.00	44.97	7.95	60.92
112C, 600 x 1200mm	nr	1.25	10.00	47.12	8.57	65.69
110T, 630 x 1050mm	nr	0.75	6.00	62.10	10.21	78.31
113T, 630 x 1350mm	nr	1.00	8.00	66.34	11.15	85.49
109V, 630 x 900mm	nr	1.25	10.00	40.69	7.60	58.29
112V, 630 x 1200mm	nr	0.75	6.00	43.31	7.40	56.71
212DG, 1200 x 1200mm	nr	1.00	8.00	38.42	6.96	53.38
212W, 1200 x 1200mm	nr	1.25	10.00	63.66	11.05	84.71
210C, 1200 x 1050mm	nr	1.50	12.00	60.04	10.81	82.85
212T, 1200 x 1200mm	nr	1.50	12.00	77.18	13.38	102.56
212CV, 1200 x 1200mm	nr	1.75	14.00	80.11	14.12	108.23
310CVC, 1770 x 1050mm	nr	1.75	14.00	107.41	18.21	139.62
413CWC, 2339 x 1350mm	nr	1.75	14.00	145.94	23.99	183.93
Standard hardwood windows without glazing bars, type						
H2N10W, 915 x 1050mm	nr	1.00	8.00	105.99	17.10	131.09
H2N13W, 915 x 1350mm	nr	1.20	9.60	110.98	18.09	138.67
H2N15W, 915 x 1500mm	nr	1.50	12.00	110.95	18.44	141.39
H210DG, 1200 x 1050mm	nr	1.50	12.00	77.21	13.38	102.59
H213W, 1200 x 1350mm	nr	1.60	12.80	129.34	21.32	163.46
H215W, 1200 x 1500mm	nr	1.70	13.60	132.45	21.91	167.96
H209CV, 1200 x 900mm	nr	1.50	12.00	160.14	25.82	197.96
H213CV, 1200 x 1300mm	nr	1.60	12.80	175.06	28.18	216.04
H309CC, 1770 x 900mm	nr	2.00	16.00	198.69	32.20	246.89
H312CC, 1770 x 1200mm	nr	2.10	16.80	216.71	35.03	268.54
H307C, 1770 x 750mm	nr	2.00	16.00	142.80	23.82	182.62
H312C, 1770 x 1200mm	nr	2.10	16.80	154.47	25.69	196.96

METAL WINDOWS

	Unit	Labour hours	Net labour (£)	Net material (£)	O'heads /profit (£)	Total (£)
L11 METAL WINDOWS						
Labour rate 8.00 per hour						
Crittalls Duralife Homelight standard galvanized steel windows, type						
NC5	nr	0.75	6.00	13.04	2.86	21.90
ND5	nr	1.25	10.00	14.89	3.73	28.62
NE13E	nr	1.00	8.00	41.09	7.36	56.45
NC13	nr	1.00	8.00	20.15	4.22	32.37
NE3	nr	1.40	11.20	51.52	9.41	72.13
NC13F	nr	1.50	12.00	46.35	8.75	67.10
ND13	nr	1.70	13.60	23.36	5.54	42.50
NC01	nr	0.75	6.00	36.66	6.40	49.06
NC5F	nr	1.00	8.00	30.25	5.74	43.99
ND5F	nr	1.25	10.00	32.15	6.32	48.47
NC05F	nr	1.00	8.00	31.16	5.87	45.03
ND1	nr	1.00	8.00	39.42	7.11	54.53
ND4F	nr	1.25	10.00	101.88	16.78	128.66
NDV5F	nr	2.50	20.00	35.50	8.32	63.82

217

	Unit	Labour hours	Net labour (£)	Net material (£)	O'heads /profit (£)	Total (£)
L20 TIMBER DOORS						
Labour rate 8.00 per hour						
Standard flush door plywood faced both sides, 35mm thick, size						
686 x 1981mm	nr	1.00	8.00	19.99	4.20	32.19
762 x 1981mm	nr	1.00	8.00	20.27	4.24	32.51
Standard flush door plywood faced both sides, 40mm thick, size						
626 x 2040mm	nr	1.20	9.60	21.05	4.60	35.25
726 x 2040mm	nr	1.20	9.60	21.29	4.63	35.52
826 x 2040mm	nr	1.20	9.60	22.24	4.78	36.62
Standard flush door sapele faced both sides, 35mm thick, size						
686 x 1981mm	nr	1.00	8.00	22.74	4.61	35.35
762 x 1981mm	nr	1.00	8.00	22.98	4.65	35.63
Standard flush door sapele faced both sides, 40mm thick, size						
626 x 2040mm	nr	1.20	9.60	23.51	4.97	38.08
726 x 2040mm	nr	1.20	9.60	23.82	5.01	38.43
826 x 2040mm	nr	1.20	9.60	24.73	5.15	39.48
Standard flush door teak faced both sides, 35mm thick, size						
686 x 1981mm	nr	1.00	8.00	40.54	7.28	55.82
762 x 1981mm	nr	1.00	8.00	41.82	7.47	57.29
Standard flush door teak faced both sides, 40mm thick, size						
626 x 2040mm	nr	1.20	9.60	44.80	8.16	62.56
726 x 2040mm	nr	1.20	9.60	40.95	7.58	58.13
826 x 2040mm	nr	1.20	9.60	45.91	8.33	63.84

TIMBER DOORS

	Unit	Labour hours	Net labour (£)	Net material (£)	O'heads /profit (£)	Total (£)
Standard flush door, half hour fire check, plywood faced both sides, 44mm thick, size						
686 x 1981mm	nr	1.20	9.60	33.64	6.49	49.73
762 x 1981mm	nr	1.20	9.60	33.64	6.49	49.73
726 x 2040mm	nr	1.20	9.60	35.30	6.73	51.63
826 x 2040mm	nr	1.20	9.60	36.82	6.96	53.38
Standard flush door, half hour fire check, sapele faced both sides, 44mm thick, size						
686 x 1981mm	nr	1.20	9.60	46.96	8.48	65.04
762 x 1981mm	nr	1.20	9.60	46.96	8.48	65.04
726 x 2040mm	nr	1.20	9.60	48.80	8.76	67.16
826 x 2040mm	nr	1.20	9.60	44.80	8.16	62.56
Framed ledged and braced door 44mm thick with 19mm matchboarding						
686 x 1981mm	nr	1.30	10.40	54.75	9.77	74.92
762 x 1981mm	nr	1.30	10.40	55.80	9.93	76.13
Framed ledged and braced door 44mm thick with 25mm matchboarding						
726 x 2040mm	nr	1.30	10.40	55.80	9.93	76.13
826 x 2040mm	nr	1.30	10.40	55.80	9.93	76.13
Door lining size 32 x 114mm with loose stops for door, size						
686 x 1981mm	nr	0.75	6.00	15.37	3.21	24.58
762 x 1981mm	nr	0.85	6.80	15.37	3.33	25.50
838 x 1981mm	nr	0.95	7.60	15.61	3.48	26.69

Linings (cont'd)	Unit	Labour hours	Net labour (£)	Net material (£)	O'heads /profit (£)	Total (£)
Door lining size 32 x 140mm with loose stops for door, size						
686 x 1981mm	nr	0.75	6.00	18.36	3.65	28.01
762 x 1981mm	nr	0.85	6.80	18.36	3.77	28.93
838 x 1981mm	nr	0.95	7.60	18.36	3.89	29.85
Door frame size 38 x 114mm rebated for door, size 762 x 1981mm	nr	1.10	8.80	24.68	0.84	34.32
Door frame size 38 x 140mm rebated for door, size 762 x 1981mm	nr	1.15	9.20	29.24	0.96	39.40

TIMBER DOORS

Linings (cont'd)	Unit	Labour hours	Net labour (£)	Net material (£)	O'heads /profit (£)	Total (£)

L30 TIMBER STAIRS

Labour rate 8.00 per hour

Standard staircases

Staircase in one flight, open
tread, 2600mm height, consisting
of 12 treads and 13 risers, width

850mm	nr	22.00	176.00	262.68	65.80	504.48
910mm	nr	22.00	176.00	278.12	68.12	522.24

Staircase in one flight, closed
tread, 2900mm height, consisting
of 14 treads and 15 risers, width

850mm	nr	23.00	184.00	475.58	98.94	758.52
914mm	nr	23.00	184.00	493.70	101.66	779.36

221

	Unit	Labour hours	Net labour (£)	Net material (£)	O'heads /profit (£)	Total (£)

L40 GENERAL GLAZING

Labour rate 8.00 per hour

Clear float glass

In wood with putty and sprigs

Under 0.15m2, thickness

3mm	m2	0.90	7.20	28.18	5.31	40.69
4mm	m2	0.90	7.20	31.94	5.87	45.01
5mm	m2	0.90	7.20	46.36	8.03	61.59
6mm	m2	1.00	8.00	51.82	8.97	68.79
10mm	m2	1.05	8.40	110.11	17.78	136.29

Over 0.15m2, thickness

3mm	m2	0.60	4.80	28.18	4.95	37.93
4mm	m2	0.60	4.80	31.94	5.51	42.25
5mm	m2	0.60	4.80	46.36	7.67	58.83
6mm	m2	0.65	5.20	51.82	8.55	65.57
10mm	m2	0.70	5.60	110.11	17.36	133.07

In wood with pinned beads

Under 0.15m2, thickness

3mm	m2	1.20	9.60	28.18	5.67	43.45
4mm	m2	1.20	9.60	31.94	6.23	47.77
5mm	m2	1.20	9.60	46.36	8.39	64.35
6mm	m2	1.35	10.80	51.82	9.39	72.01
10mm	m2	1.50	12.00	110.11	18.32	140.43

Over 0.15m2, thickness

3mm	m2	0.80	6.40	28.18	5.19	39.77
4mm	m2	0.80	6.40	31.94	5.75	44.09
5mm	m2	0.80	6.40	46.36	7.91	60.67
6mm	m2	0.90	7.20	51.82	8.85	67.87
10mm	m2	1.00	8.00	110.11	17.72	135.83

In wood with screwed beads

Under 0.15m2, thickness

3mm	m2	1.65	13.20	28.18	6.21	47.59
4mm	m2	1.65	13.20	31.94	6.77	51.91
5mm	m2	1.65	13.20	46.36	8.93	68.49

GENERAL GLAZING

	Unit	Labour hours	Net labour (£)	Net material (£)	O'heads /profit (£)	Total (£)
6mm	m2	1.80	14.40	51.82	9.93	76.15
10mm	m2	1.95	15.60	110.11	18.86	144.57
Over 0.15m2, thickness						
3mm	m2	1.10	8.80	28.18	5.55	42.53
4mm	m2	1.10	8.80	31.94	6.11	46.85
5mm	m2	1.10	8.80	46.36	8.27	63.43
6mm	m2	1.20	9.60	51.82	9.21	70.63
10mm	m2	1.30	10.40	110.11	18.08	138.59
In metal with putty						
Under 0.15m2, thickness						
3mm	m2	1.20	9.60	28.18	5.67	43.45
4mm	m2	1.20	9.60	31.94	6.23	47.77
5mm	m2	1.20	9.60	46.36	8.39	64.35
6mm	m2	1.35	10.80	51.82	9.39	72.01
10mm	m2	1.50	12.00	110.11	18.32	140.43
Over 0.15m2, thickness						
3mm	m2	0.80	6.40	28.18	5.19	39.77
4mm	m2	0.80	6.40	31.94	5.75	44.09
5mm	m2	0.80	6.40	46.36	7.91	60.67
6mm	m2	0.90	7.20	51.82	8.85	67.87
10mm	m2	1.00	8.00	110.11	17.72	135.83
In metal with clipped beads						
Under 0.15m2, thickness						
3mm	m2	1.50	12.00	28.18	6.03	46.21
4mm	m2	1.50	12.00	31.94	6.59	50.53
5mm	m2	1.50	12.00	46.36	8.75	67.11
6mm	m2	1.65	13.20	51.82	9.75	74.77
10mm	m2	1.80	14.40	110.11	18.68	143.19
Over 0.15m2, thickness						
3mm	m2	1.00	8.00	28.18	5.43	41.61
4mm	m2	1.00	8.00	31.94	5.99	45.93
5mm	m2	1.00	8.00	46.36	8.15	62.51
6mm	m2	1.10	8.80	51.82	9.09	69.71
10mm	m2	1.20	9.60	110.11	17.96	137.67

RATES FOR MEASURED WORK

	Unit	Labour hours	Net labour (£)	Net material (£)	O'heads /profit (£)	Total (£)
White patterned						
In wood with putty and sprigs						
Under 0.15m2, thickness						
4mm	m2	0.90	7.20	23.63	4.62	35.45
5mm	m2	0.90	7.20	33.16	6.05	46.41
6mm	m2	1.00	8.00	42.32	7.55	57.87
Over 0.15m2, thickness						
3mm	m2	0.60	4.80	23.63	4.26	32.69
5mm	m2	0.60	4.80	33.16	5.69	43.65
6mm	m2	0.65	5.20	42.32	7.13	54.65
In wood with pinned beads						
Under 0.15m2, thickness						
4mm	m2	1.20	9.60	23.63	4.98	38.21
5mm	m2	1.20	9.60	33.16	6.41	49.17
6mm	m2	1.35	10.80	42.32	7.97	61.09
Over 0.15m2, thickness						
4mm	m2	0.80	6.40	23.63	4.50	34.53
5mm	m2	0.80	6.40	33.16	5.93	45.49
6mm	m2	0.90	7.20	42.32	7.43	56.95
In wood with screwed beads						
Under 0.15m2, thickness						
4mm	m2	1.65	13.20	23.63	5.52	42.35
5mm	m2	1.65	13.20	33.16	6.95	53.31
6mm	m2	1.80	14.40	42.32	8.51	65.23
Over 0.15m2, thickness						
4mm	m2	1.10	8.80	23.63	4.86	37.29
5mm	m2	1.10	8.80	33.16	6.29	48.25
6mm	m2	1.20	9.60	42.32	7.79	59.71

M
SURFACE FINISHES

MATERIAL COSTS

	Unit	Material supply (£)	Material waste (%)	Total (£)
MATERIAL COSTS				
Sand cement/granolithic screeds				
Cement	t	82.75	10.0	91.03
Sand	m3	15.50	10.0	17.05
Granolithic chippings	t	21.00	10.0	23.10
Liquid hardening agent	25l	26.08	10.0	28.69
Oil repellent agent	25l	113.92	10.0	125.31
Resin/Latex flooring				
Febflor	25kg	19.08	10.0	20.99
Nicobond latex screed mix	each	17.59	10.0	19.35
Vermiculite	m3	83.16	10.0	91.48
Plastered/rendered/roughcast coatings				
Plaster				
browning	t	137.56	5.0	144.44
bonding	t	145.15	5.0	152.41
finishings	t	126.62	5.0	132.95
plasterboard finish	t	112.81	5.0	118.45
universal	t	187.07	5.0	196.42
multi finish	t	119.71	5.0	125.70
metal lathing	t	157.79	5.0	165.68
renovating finish	t	168.03	5.0	176.43
Angle beads				
standard	m	0.44	10.0	0.48
Thin coat				
3mm	m	0.64	10.0	0.70
6mm	m	0.64	10.0	0.70

RATES FOR MEASURED WORK

	Unit	Material supply (£)	Material waste (%)	Total (£)
drywall corner	m	0. 40	10. 0	0. 44
standard stop	m	0. 61	10. 0	0. 67
plasterboard edging board	m	1. 72	10. 0	1. 89
Cullamix	50kg	19. 24	5. 0	20. 20
Alpine	50kg	29. 57	5. 0	31. 05
Sandtex primer	5l	14. 17	5. 0	14. 88

Metal mesh lathing

Expamet metal lath

BB263	m2	2. 55	10. 0	2. 80
BB264	m2	2. 97	10. 0	3. 27
94G	m2	4. 56	10. 0	5. 02
269	m2	3. 21	10. 0	3. 53

Quarry/ceramic tiling

Quarry tiles, plain, red

150 x 150 x 12.5mm	m2	15. 95	5. 0	16. 75
200 x 200 x 19mm	m2	22. 02	5. 0	23. 12

Quarry tiles, plain, heatherbrown

150 x 150 x 12.5mm	m2	16. 19	5. 0	17. 00
194 x 194 x 12.5mm	m2	15. 33	5. 0	16. 10

Ceramic wall tiles
152 x 152 x 6.5mm
price group

A	m2	15. 69	5. 0	16. 47
B	m2	15. 81	5. 0	16. 60
B1	m2	17. 50	5. 0	18. 38
C	m2	19. 62	5. 0	20. 60

MATERIAL COSTS

	Unit	Material supply (£)	Material waste (%)	Total (£)
Ceramic wall tiles 203 x 152 x 6.5mm price group				
A	m2	18.31	5.0	19.23
B	m2	19.21	5.0	20.17
C	m2	20.98	5.0	22.03
Ceramic studio wall tiles 200 x 200 x 6.5mm price group				
A	m2	20.73	5.0	21.77
B	m2	23.43	5.0	24.60
C	m2	25.76	5.0	27.05
D	m2	26.12	5.0	27.43
Ceramic Studio wall tile 200 x 200 x 7mm price group				
A	m2	20.73	5.0	21.77
B	m2	23.43	5.0	24.60
Ceramic floor tiles Dover 150 x 150 x 9mm				
Grey steel white	m2	16.38	5.0	17.20
Stone, rockface, russet blend	m2	13.65	5.0	14.33
Black	m2	14.06	5.0	14.76
Red	m2	11.56	5.0	12.14
Rubber/plastics/cork/lino/carpet				
Plastic tiles				
Econoflex series 2: 2mm thick	m2	2.67	5.0	2.80
Econoflex series 4: 2mm thick	m2	3.09	5.0	3.24
Econoflex series 2: 2.5mm thick	m2	2.95	5.0	3.10
Econoflex series 4: 2.5mm thick	m2	3.34	5.0	3.51
Marleyflex 2mm thick	m2	3.32	5.0	3.49
Marleyflex 2.5mm thick	m2	3.98	5.0	4.18
Vylon 2mm thick	m2	3.47	5.0	3.64
Travertone 2.5mm thick	m2	4.28	5.0	4.49

RATES FOR MEASURED WORK

Plastic tiles (cont'd)	Unit	Material supply (£)	Material waste (%)	Total (£)
Sculpture 1.8mm thick	m2	5.32	5.0	5.59
Onyx/Classic granite 3.2mm thick	m2	11.45	5.0	12.02
Piazza 2.5mm thick	m2	22.40	5.0	23.52
Medina 3mm thick	m2	6.40	5.0	6.72

Vinyl polyflex tiles

Premier: 2mm thick	m2	5.25	5.0	5.51
XL: 2mm thick	m2	4.86	5.0	5.10
Super XL: 2.5mm thick	m2	6.04	5.0	6.34
Super XL: 3mm thick	m2	7.25	5.0	7.61
Antistatic: 2mm thick	m2	7.34	5.0	7.71
Static conductive: 2mm thick	m2	13.40	5.0	14.07

Vinyl sheeting

Marleyflor, 2mm	m2	4.24	5.0	4.45
HD vinyl sheet, 2mm	m2	5.43	5.0	5.70
HD vinyl sheet, 2.5mm	m2	6.28	5.0	6.59
HD acoustic sheet	m2	6.20	5.0	6.51
HD Hitech, 2mm	m2	7.44	5.0	7.81

MATERIAL COSTS

	Unit	Material supply (£)	Material waste (%)	Total (£)
Decorative papers				
Lining paper	roll	1.00	20.0	1.20
Washable paper	roll	2.50	20.0	3.00
Vinyl paper	roll	5.00	20.0	6.00
Embossed paper	roll	4.00	20.0	4.80
Embossed paper, heavy	roll	6.00	20.0	7.20
Textured paper	roll	4.00	20.0	4.80
Hessian paper	m2	3.00	10.0	3.30
Hessian paper, heavy	m2	4.00	10.0	4.40
Textile wall coverings	m2	8.00	10.0	8.80
Textile wall coverings	m2	10.00	10.0	11.00
Suede coverings	m2	10.00	10.0	11.00
Painting				
Primers				
Emulsion	5l	15.00	5.0	15.75
Primer sealer	5l	16.38	5.0	17.20
Acrylic primer	5l	13.81	5.0	14.50
Zinc phosphate	5l	14.81	5.0	15.55
Quick drying metal	5l	16.28	5.0	17.09
Calcium plumbate	5l	26.68	5.0	28.01

231

RATES FOR MEASURED WORK

	Unit	Material supply (£)	Material waste (%)	Total (£)
Emulsion				
Matt white	5l	16. 54	5. 0	17. 37
Matt coloured	5l	19. 62	5. 0	20. 60
Silk brilliant white	5l	17. 20	5. 0	18. 06
Oil based				
Undercoat white	5l	19. 62	5. 0	20. 60
Gloss white	5l	19. 62	5. 0	20. 60
Gloss brilliant white	5l	19. 62	5. 0	20. 60
Gloss coloured	5l	24. 80	5. 0	26. 04
Eggshell white	5l	20. 73	5. 0	21. 77
Eggshell brilliant white	5l	20. 73	5. 0	21. 77

	Unit	Labour hours	Net labour (£)	Net material (£)	O'heads /profit (£)	Total (£)
M10 SAND CEMENT/GRANO/SCREEDS						
Labour rate 10.33 per hour						
Cement and sand (1:3) beds screeded finish						
Floors, level and to falls not exceeding 15 degrees from horizontal						
25mm thick						
over 300mm wide	m2	0.21	2.17	1.51	0.55	4.23
not exceeding 300mm wide	m	0.08	0.83	0.45	0.19	1.47
32mm thick						
over 300mm wide	m2	0.23	2.38	1.93	0.65	4.96
not exceeding 300mm wide	m	0.75	7.75	0.58	1.25	9.58
38mm thick						
over 300mm wide	m2	0.25	2.58	2.29	0.73	5.60
not exceeding 300mm wide	m	0.10	1.03	0.68	0.26	1.97
50mm thick						
over 300mm wide	m2	0.29	3.00	3.01	0.90	6.91
not exceeding 300mm wide	m	0.12	1.24	0.90	0.32	2.46
63mm thick						
over 300mm wide	m2	0.33	3.41	3.78	1.08	8.27
not exceeding 300mm wide	m	0.13	1.34	1.15	0.37	2.86
Increase the labour rate element of all working in boiler rooms or compartments not exceeding 4m2 by 15%						
Linings to channels, to falls, square arrises and junctions						
25mm thick						
150mm girth	m	0.35	3.62	0.23	0.58	4.43
250mm girth	m	0.58	5.99	0.39	0.96	7.34
450mm girth	m	1.04	10.74	0.68	1.71	13.13

RATES FOR MEASURED WORK

Cement and sand screeds (cont'd)	Unit	Labour hours	Net labour (£)	Net material (£)	O'heads /profit (£)	Total (£)
32mm thick						
150mm girth	m	0.40	4.13	0.29	0.66	5.08
250mm girth	m	0.62	6.40	0.48	1.03	7.91
450mm girth	m	1.11	11.47	0.87	1.85	14.19
38mm thick						
250mm girth	m	0.66	6.82	0.58	1.11	8.51
450mm girth	m	1.19	12.29	1.13	2.01	15.43
600mm girth	m	1.58	16.32	1.37	2.65	20.34
50mm thick						
250mm girth	m	0.69	7.13	0.75	1.18	9.06
450mm girth	m	1.24	12.81	1.35	2.12	16.28
600mm girth	m	1.66	17.15	1.80	2.84	21.79
63mm thick						
450mm girth	m	0.74	7.64	1.71	1.40	10.75
600mm girth	m	1.29	13.33	2.37	2.35	18.06
700mm girth	m	1.71	17.66	2.65	3.05	23.36
Landings						
25mm thick						
over 300mm wide	m2	0.25	2.58	2.16	0.71	5.45
not exceeding 300mm wide	m	0.00	0.00	0.45	0.07	0.52
32mm thick						
over 300mm wide	m2	0.26	2.69	1.93	0.69	5.31
not exceeding 300mm wide	m	0.10	1.03	0.58	0.24	1.85
38mm thick						
over 300mm wide	m2	0.28	2.89	2.29	0.78	5.96
not exceeding 300mm wide	m	0.11	1.14	0.69	0.27	2.10
50mm thick						
over 300mm wide	m2	0.32	3.31	3.61	1.04	7.96
not exceeding 300mm wide	m	0.13	1.34	0.91	0.34	2.59
63mm thick						
over 300mm wide	m2	0.36	3.72	3.80	1.13	8.65
not exceeding 300mm wide	m	0.14	1.45	1.64	0.46	3.55

SAND CEMENT/GRANOLITHIC

	Unit	Labour hours	Net labour (£)	Net material (£)	O'heads /profit (£)	Total (£)
Treads						
25mm thick, width						
275mm	m	0.21	2.17	0.42	0.39	2.98
350mm	m	0.08	0.83	0.54	0.21	1.58
38mm thick, width						
275mm	m	0.25	2.58	0.63	0.48	3.69
350mm	m	0.10	1.03	0.82	0.28	2.13
50mm thick, width						
275mm	m	0.28	2.89	0.83	0.56	4.28
350mm	m	0.11	1.14	1.06	0.33	2.53
Add to the previous prices for floated finish	m2	0.15	1.55	0.00	0.04	1.59
Add to the previous prices for trowelled finish	m2	0.23	2.38	0.00	0.06	2.44
Additives and surface dressings						
One coat liquid hardening agent	m2	0.20	2.07	0.29	0.35	2.71
Two coats liquid hardening agent	m2	0.18	1.86	0.57	0.36	2.79
One coat oil repellent agent	m2	0.10	1.03	0.63	0.25	1.91
Granolithic, cement and granite chippings (1:2.5), steel trowelled smooth						
Floors, level and to falls not exceeding 15 degrees from horizontal						
25mm thick						
over 300mm wide	m2	0.38	3.93	2.83	1.01	7.77
not exceeding 300mm wide	m	0.15	1.55	0.82	0.36	2.73
32mm thick						
over 300mm wide	m2	0.42	4.34	3.50	1.18	9.02
not exceeding 300mm wide	m	1.34	13.84	1.05	2.23	17.12

RATES FOR MEASURED WORK

Granolithic (cont'd)	Unit	Labour hours	Net labour (£)	Net material (£)	O'heads /profit (£)	Total (£)
38mm thick						
over 300mm wide	m2	0.45	4.65	4.15	1.32	10.12
not exceeding 300mm wide	m	0.18	1.86	1.25	0.47	3.58
50mm thick						
over 300mm wide	m2	0.48	4.96	5.46	1.56	11.98
not exceeding 300mm wide	m	0.19	1.96	1.64	0.54	4.14
63mm thick						
over 300mm wide	m2	0.52	5.37	6.88	1.84	14.09
not exceeding 300mm wide	m	0.21	2.17	2.06	0.63	4.86

Increase the labour element of all
items for working in boiler rooms
or compartments not exceeding
4m2 by 15%

Linings to channels, to falls,
square arrises and junctions

	Unit	Labour hours	Net labour (£)	Net material (£)	O'heads /profit (£)	Total (£)
25mm thick						
150mm girth	m	0.35	3.62	0.41	0.60	4.63
250mm girth	m	0.58	5.99	0.69	1.00	7.68
450mm girth	m	1.04	10.74	1.23	1.80	13.77
32mm thick						
150mm girth	m	0.40	4.13	0.53	0.70	5.36
250mm girth	m	0.88	9.09	0.87	1.49	11.45
450mm girth	m	1.30	13.43	1.57	2.25	17.25
38mm thick						
250mm girth	m	0.66	6.82	1.03	1.18	9.03
450mm girth	m	1.19	12.29	1.86	2.12	16.27
600mm girth	m	1.58	16.32	2.48	2.82	21.62
50mm thick						
250mm girth	m	0.69	7.13	1.37	1.27	9.78
450mm girth	m	1.24	12.81	2.45	2.29	17.55
600mm girth	m	1.66	17.15	3.27	3.06	23.48
63mm thick						
250mm girth	m	0.82	8.47	1.72	1.53	11.72
450mm girth	m	1.56	16.11	3.09	2.88	22.08
600mm girth	m	1.98	20.45	4.13	3.69	28.27

SAND CEMENT/GRANOLITHIC

	Unit	Labour hours	Net labour (£)	Net material (£)	O'heads /profit (£)	Total (£)
Landings						
25mm thick						
over 300mm wide	m2	0.42	4.34	2.73	1.06	8.13
not exceeding 300mm wide	m	0.17	1.76	0.82	0.39	2.97
32mm thick						
over 300mm wide	m2	0.46	4.75	3.49	1.24	9.48
not exceeding 300mm wide	m	0.18	1.86	1.05	0.44	3.35
38mm thick						
over 300mm wide	m2	0.50	5.17	4.15	1.40	10.72
not exceeding 300mm wide	m	0.20	2.07	1.26	0.50	3.83
50mm thick						
over 300mm wide	m2	0.53	5.47	5.46	1.64	12.57
not exceeding 300mm wide	m	0.21	2.17	1.64	0.57	4.38
63mm thick						
over 300mm wide	m2	0.56	5.78	6.87	1.90	14.55
not exceeding 300mm wide	m	0.22	2.27	2.06	0.65	4.98
Treads						
25mm thick, width						
275mm	m	0.28	2.89	0.79	0.55	4.23
350mm	m	0.35	3.62	0.96	0.69	5.27
32mm thick, width						
275mm	m	0.32	3.31	0.97	0.64	4.92
350mm	m	0.39	4.03	1.23	0.79	6.05
38mm thick, width						
275mm	m	0.36	3.72	1.15	0.73	5.60
350mm	m	0.43	4.44	1.44	0.88	6.76
50mm thick, width						
275mm	m	0.40	4.13	1.50	0.84	6.47
350mm	m	0.47	4.86	1.91	1.02	7.79
63mm thick, width						
275mm	m	0.45	4.65	1.89	0.98	7.52
350mm	m	0.52	5.37	2.40	1.17	8.94

	Unit	Labour hours	Net labour (£)	Net material (£)	O'heads /profit (£)	Total (£)

M12 RESIN/LATEX FLOORING

Labour rate 10.33 per hour

Self-levelling epoxy floor topping Feblor or similar

Floors, level and to floors not exceeding 15 degrees from horizontal, 1.5mm thick

	Unit	Labour hours	Net labour (£)	Net material (£)	O'heads /profit (£)	Total (£)
over 300mm wide	m2	0. 35	3. 62	1. 31	0. 12	5. 05
not exceeding 300mm wide	m	0. 14	1. 45	0. 39	0. 05	1. 89

Increase the labour element of all items for working in compartments not exceeding 4m2 by 15%

Making good around pipes and the like

	Unit	Labour hours	Net labour (£)	Net material (£)	O'heads /profit (£)	Total (£)
not exceeding 0.3m girth	nr	0. 08	0. 83	0. 00	0. 12	0. 95
0.3m to 1m girth	nr	0. 10	1. 03	0. 00	0. 15	1. 18
1m to 2m girth	nr	0. 12	1. 24	0. 00	0. 19	1. 43

Additives and surface dressings

	Unit	Labour hours	Net labour (£)	Net material (£)	O'heads /profit (£)	Total (£)
silicon carbide granules sprinkled on	m2	0. 08	0. 83	0. 50	0. 20	1. 53

	Unit	Labour hours	Net labour (£)	Net material (£)	O'heads /profit (£)	Total (£)
M20 PLASTERED/RENDERED/ROUGHCAST COATING						

Labour rate 10.33 per hour

INTERNAL WORK

Premixed lightweight plaster, Carlite, 11mm bonding, 2mm finish

To low suction concrete, brick or block

Walls

over 300mm wide	m2	0.44	4.55	1.50	0.91	6.96
not exceeding 300mm wide	m	0.18	1.86	0.45	0.35	2.66

Curved walls

over 300mm wide	m2	0.70	7.23	1.50	1.31	10.04
not exceeding 300mm wide	m	0.28	2.89	0.45	0.50	3.84

Ceilings

over 300mm wide	m2	0.57	5.89	1.50	1.11	8.50
not exceeding 300mm wide	m	0.23	2.38	0.45	0.42	3.25

Sides, soffits or tops of isolated beams

over 300mm wide	m2	0.70	7.23	1.50	1.31	10.04
not exceeding 300mm wide	m	0.28	2.89	0.45	0.50	3.84

Sides of isolated columns

over 300mm wide	m2	0.48	4.96	1.50	0.97	7.43
not exceeding 300mm wide	m	0.19	1.96	0.45	0.36	2.77

Increase the labour element for working in staircase areas not exceeding 4m2 by 15%

Plasterwork (cont'd)	Unit	Labour hours	Net labour (£)	Net material (£)	O'heads /profit (£)	Total (£)
Increase the labour element of ceilings and beams for work over 3.5m high by 10%						
Plaster, 5mm two coat Thistle board finish, to plasterboard or concrete						
Walls						
over 300mm wide	m2	0.40	4.13	0.60	0.71	5.44
not exceeding 300mm wide	m	0.26	2.69	0.18	0.43	3.30
Curved walls						
over 300mm wide	m2	0.63	6.51	0.60	1.07	8.18
not exceeding 300mm wide	m	0.23	2.38	0.18	0.38	2.94
Ceilings						
over 300mm wide	m2	0.51	5.27	0.60	0.88	6.75
not exceeding 300mm wide	m	0.21	2.17	0.18	0.35	2.70
Sides, soffits or tops of isolated beams						
over 300mm wide	m2	0.63	6.51	0.60	1.07	8.18
not exceeding 300mm wide	m	0.23	2.38	0.18	0.38	2.94
Sides of isolated columns						
over 300mm wide	m2	0.43	4.44	0.60	0.76	5.80
not exceeding 300mm wide	m	0.17	1.76	0.18	0.29	2.23

Increase the labour element of all items for working in staircase areas or compartments not exceeding 4m2 by 15%

Increase the labour element of ceilings and beams for work over 3.5m high by 10%

PLASTERED/RENDERED/ROUGHCAST

	Unit	Labour hours	Net labour (£)	Net material (£)	O'heads /profit (£)	Total (£)
Cement and sand (1:3) backings, floated finish						
To walls 13mm thick						
over 300mm wide	m2	0.35	3.62	0.80	0.66	5.08
not exceeding 300mm wide	m	0.00	0.00	0.24	0.04	0.28
To walls 19mm thick						
over 300mm wide	m2	0.39	4.03	1.17	0.78	5.98
not exceeding 300mm wide	m	0.00	0.00	0.36	0.05	0.41
To walls 25mm thick						
over 300mm wide	m2	0.43	4.44	0.54	0.75	5.73
not exceeding 300mm wide	m	0.00	0.00	0.46	0.07	0.53
Increase the labour element of all items for working in compartments not exceedingg 4m2 by 15%						
Sides of isolated columns						
13mm thick						
over 300mm wide	m2	0.39	4.03	0.80	0.72	5.55
not exceeding 300mm wide	m	0.00	0.00	0.24	0.04	0.28
19mm thick						
over 300mm wide	m2	0.43	4.44	1.17	0.84	6.45
not exceeding 300mm wide	m	0.17	1.76	0.36	0.05	2.17
25mm thick						
over 300mm wide	m2	0.47	4.86	0.54	0.81	6.21
not exceeding 300mm wide	m	0.19	1.96	0.46	0.36	2.78

Cement and sand backings (cont'd)	Unit	Labour hours	Net labour (£)	Net material (£)	O'heads /profit (£)	Total (£)
Increase the labour element of all items for working in compartments not exceeding 4m2 by 15%						
Additives and surface dressings						
One coat liquid hardening agent	m2	0.20	2.07	0.28	0.35	2.70
Two coats liquid hardening agent	m2	0.16	1.65	0.56	0.33	2.54
One coat oil repellent agent	m2	0.10	1.03	0.63	0.25	1.91
EXTERNAL WORK						
Render, cement and sand (1:3), 13mm one coat work, plain face						
Walls						
over 300mm wide	m2	0.35	3.62	0.80	0.66	5.08
not exceeding 300mm wide	m	0.14	1.45	0.24	0.25	1.94
Curved walls						
over 300mm wide	m2	0.56	5.78	0.80	0.99	7.57
not exceeding 300mm wide	m	0.20	2.07	0.24	0.35	2.66
Sides of isolated columns						
over 300mm wide	m2	0.39	4.03	0.80	0.72	5.55
not exceeding 300mm wide	m	0.16	1.65	0.24	0.28	2.17
Add to the previous prices for						
integral waterproofer	m2	0.05	0.52	0.33	0.13	0.98
roughcast face	m2	0.25	2.58	0.00	0.39	2.97
pebble dash finish	m2	0.30	3.10	0.22	0.50	3.82
priming coat and bonding one coat of Febond	m2	0.20	2.07	3.61	0.85	6.53
cement and sand spatterdash for key	m2	0.16	1.65	0.42	0.31	2.38

PLASTERED/RENDERED/ROUGHCAST

	Unit	Labour hours	Net labour (£)	Net material (£)	O'heads /profit (£)	Total (£)
Tyrolean render, Cullamix, three applications						
Walls						
over 300mm wide	m2	0.30	3.10	5.98	1.36	10.44
not exceeding 300mm wide	m	0.12	1.24	1.99	0.08	3.31
Curved walls						
over 300mm wide	m2	0.33	3.41	5.98	1.41	10.80
not exceeding 300mm wide	m	0.13	1.34	1.99	0.50	3.83
Sides of isolated columns						
over 300mm wide	m2	0.36	3.72	5.98	1.45	11.16
not exceeding 300mm wide	m	0.15	1.55	1.99	0.53	4.07
Add to the previous prices for						
rubber finish	m2	0.15	1.55	0.00	0.23	1.78
Cemprover PVA/Cullamix scrubcoat key	m	0.10	1.03	0.37	0.21	1.61
Dragged rendering, Alpine, one coat						
Walls						
over 300mm wide	m2	0.20	2.07	2.70	0.72	5.49
not exceeding 300mm wide	m	0.08	0.83	0.80	0.24	1.87
Curved walls						
over 300mm wide	m2	0.30	3.10	2.70	0.87	6.67
not exceeding 300mm wide	m	0.12	1.24	0.80	0.31	2.35
Sides of isolated columns						
over 300mm wide	m2	0.25	2.58	2.70	0.79	6.07
not exceeding 300mm wide	m	0.10	1.03	0.80	0.27	2.10
Add to previous prices for priming coat of Sandtex Matt	m2	0.10	1.03	0.48	0.23	1.74

	Unit	Labour hours	Net labour (£)	Net material (£)	O'heads /profit (£)	Total (£)

Textured surface coating, Cullamix 7mm thick, sprayed finish

Walls

over 300mm wide	m2	0.10	1.03	2.12	0.47	3.62
not exceeding 300mm wide	m	0.15	1.55	0.64	0.33	2.52

Curved walls

over 300mm wide	m2	0.11	1.14	2.12	0.49	3.75
not exceeding 300mm wide	m	0.17	1.76	0.64	0.36	2.76

Sides of isolated columns

over 300mm wide	m2	0.12	1.24	2.12	0.50	3.86
not exceeding 300mm wide	m	1.19	12.29	0.64	1.94	14.87

Add to the previous prices for the work to existing surfaces including fungicidal wash, filling of cracks and stabilizing solution	m2	0.20	2.07	0.56	0.39	3.02
Add to the previous prices for priming coat of Sandtex Matt	m2	0.10	1.03	0.43	0.22	1.68

METAL MESH LATHING

	Unit	Labour hours	Net labour (£)	Net material (£)	O'heads /profit (£)	Total (£)

M30 METAL MESH LATHING

Labour rate 10.33 per hour

INTERNAL WORK

Metal lathing, galvanized Expamet, 700 x 2500mm to timber with galvanized nails

Reference BB263

Walls

	Unit	Labour hours	Net labour (£)	Net material (£)	O'heads /profit (£)	Total (£)
over 300mm wide	m2	0.17	1.76	2.83	0.69	5.28
not exceeding 300mm wide	m	0.07	0.72	0.82	0.23	1.77

Ceilings

over 300mm wide	m2	0.20	2.07	2.83	0.73	5.64
not exceeding 300mm wide	m	0.08	0.83	0.82	0.25	1.90

Sides, soffits or tops of isolated beams

over 300mm wide	m2	0.25	2.58	2.83	0.81	6.22
not exceeding 300mm wide	m	0.10	1.03	0.82	0.28	2.13

Sides of isolated columns

over 300mm wide	m2	0.25	2.58	2.83	0.81	6.22
not exceeding 300mm wide	m	0.10	1.03	0.82	0.05	1.90

Add to the previous prices for fixing with tying wire to supporting metalwork	m2	0.04	0.41	0.00	0.06	0.47

Increase the labour element of all items for working in staircase areas or compartments not exceeding 4m2 by 15%

Lathing (cont'd)	Unit	Labour hours	Net labour (£)	Net material (£)	O'heads /profit (£)	Total (£)
Increase the labour element of ceilings and beams for work over 3.5m high by 10%						

Reference BB265

Walls

	Unit	Labour hours	Net labour (£)	Net material (£)	O'heads /profit (£)	Total (£)
over 300mm wide	m2	0.17	1.76	2.61	0.66	5.03
not exceeding 300mm wide	m	0.07	0.72	2.61	0.50	3.83

Ceilings

	Unit	Labour hours	Net labour (£)	Net material (£)	O'heads /profit (£)	Total (£)
over 300mm wide	m2	0.20	2.07	2.61	0.70	5.38
not exceeding 300mm wide	m	0.08	0.83	2.61	0.52	3.96

Sides, soffits or tops of isolated beams

	Unit	Labour hours	Net labour (£)	Net material (£)	O'heads /profit (£)	Total (£)
over 300mm wide	m2	0.25	2.58	2.61	0.78	5.97
not exceeding 300mm wide	m	0.10	1.03	2.61	0.55	4.19

Sides of isolated columns

	Unit	Labour hours	Net labour (£)	Net material (£)	O'heads /profit (£)	Total (£)
over 300mm wide	m2	0.25	2.58	2.61	0.78	5.97
not exceeding 300mm wide	m	0.10	1.03	2.61	0.09	3.73

	Unit	Labour hours	Net labour (£)	Net material (£)	O'heads /profit (£)	Total (£)
Add to the previous prices for fixing with tying wire to supporting metalwork	m2	0.04	0.41	0.00	0.01	0.42

Increase the labour element of all items for working in staircase areas or compartments not exceeding 4m2 by 15%

Increase the labour element of ceilings and beams for work over 3.5m high by 10%

METAL MESH LATHING

	Unit	Labour hours	Net labour (£)	Net material (£)	O'heads /profit (£)	Total (£)
Reference 95S						
Walls						
over 300mm wide	m2	0.23	2.38	7.86	1.54	11.78
not exceeding 300mm wide	m	0.09	0.93	2.50	0.51	3.94
Ceilings						
over 300mm wide	m2	0.26	2.69	7.86	1.58	12.13
not exceeding 300mm wide	m	0.10	1.03	2.50	0.53	4.06
Sides, soffits or tops of isolated beams						
over 300mm wide	m2	0.31	3.20	7.86	1.66	12.72
not exceeding 300mm wide	m	0.12	1.24	2.50	0.56	4.30
Sides of isolated columns						
over 300mm wide	m2	0.31	3.20	7.86	1.66	12.72
not exceeding 300mm wide	m	0.12	1.24	2.50	0.56	4.30
Add to the previous prices for fixing with tying wire to supporting metalwork	m2	0.04	0.41	0.00	0.06	0.47

Increase the labour element
of all items for working in
staircase areas or compartments
not exceeding 4m2 by 15%

Increase the labour element of
ceilings and beams for work over
3.5m high by 10%

	Unit	Labour hours	Net labour (£)	Net material (£)	O'heads /profit (£)	Total (£)

M40 QUARRY/CERAMIC TILING

Labour rate 10.33 per hour

QUARRY TILES

Red clay quarry floor tiles, bedding in 12mm cement mortar (1:3), butt joints straight both ways

Floors, level and to falls not exceeding 15 degrees to the horizontal

150 x 150 x 12.5mm thick

over 300mm wide	m2	0.90	9.30	16.75	3.91	29.96
not exceeding 300mm wide	m	0.36	3.72	5.07	1.32	10.11

200 x 200 x 19mm thick

over 300mm wide	m2	0.80	8.26	23.12	4.71	36.09
not exceeding 300mm wide	m	0.32	3.31	7.00	1.55	11.86

Increase the labour element of all items for working in boiler rooms or compartments not exceeding 4m2 by 15%

Landings

Landings 150 x 150 x 12.5mm thick

over 300mm wide	m2	0.99	10.23	16.75	4.05	31.03
not exceeding 300mm wide	m	0.40	4.13	5.07	1.38	10.58

Landings 200 x 200 x 19mm thick

over 300mm wide	m2	0.88	9.09	23.12	4.83	37.04
not exceeding 300mm wide	m	0.36	3.72	7.00	1.61	12.33

QUARRY/CERAMIC TILING

	Unit	Labour hours	Net labour (£)	Net material (£)	O'heads /profit (£)	Total (£)
Treads 150 x 150 x 12.5mm thick, width						
300mm	m	0.56	5.78	3.46	1.39	10.63
Treads 200 x 200 x 19.5mm thick, width						
225mm	m	0.50	5.17	4.20	1.41	10.78
Risers 150 x 150 x 12.5mm thick, height						
150mm	m	0.41	4.24	1.53	0.87	6.64
150mm, undercut	m	0.46	4.75	1.53	0.94	7.22
200mm	m	0.54	5.58	3.06	1.30	9.94
200mm, undercut	m	0.59	6.09	3.06	1.37	10.52
Strings or aprons, rounded top edge						
150 x 150 x 12.5mm thick, width						
275mm	m	0.74	7.64	3.46	1.66	12.76
Skirting, rounded top edge, coved junction to paving						
12.5mm thick, height						
150mm	m	0.25	2.58	1.53	0.62	4.73
Extra over for						
end	nr	0.05	0.52	1.45	0.30	2.27
angle	nr	0.07	0.72	1.45	0.33	2.50
ramp	nr	0.09	0.93	0.00	0.14	1.07
19mm thick, height						
200mm	nr	0.30	3.10	4.20	1.09	8.40
Extra over for						
end	nr	0.06	0.62	1.45	0.31	2.38
angle	nr	0.08	0.83	1.45	0.34	2.62
ramp	nr	0.10	1.03	0.00	0.15	1.18

249

RATES FOR MEASURED WORK

	Unit	Labour hours	Net labour (£)	Net material (£)	O'heads /profit (£)	Total (£)
CERAMIC TILES						
Glazed ceramic studio tiles, Pilkington or similar, fixing with thin bed adhesive, pointing with matching grout						
To walls, 152 x 152 x 5.6mm thick						
Over 300mm wide, price group						
Studio A	m2	0.95	9.81	16.47	3.94	30.22
Studio B	m2	0.95	9.81	16.47	3.94	30.22
Studio B1	m2	0.95	9.81	18.38	4.23	32.42
Studio C	m2	0.95	9.81	20.60	4.56	34.97
Not exceeding 300mm wide, price group						
Studio A	m	0.38	3.93	4.99	1.34	10.26
Studio B	m	0.38	3.93	4.99	1.34	10.26
Studio B1	m	0.38	3.93	5.57	1.43	10.93
Studio C	m	0.38	3.93	6.24	1.53	11.70
To walls, 152 x 152 x 8mm thick						
Over 300mm wide, price group						
Studio A	m2	1.00	10.33	19.23	4.43	33.99
Studio B	m2	1.00	10.33	20.17	4.58	35.08
Studio C	m2	1.00	10.33	22.03	4.85	37.21
Not exceeding 300mm wide, price group						
Studio A	m	0.30	3.10	5.82	1.34	10.26
Studio B	m	0.30	3.10	6.11	1.38	10.59
Studio C	m	0.30	3.10	6.68	1.47	11.25
To walls, 200 x 200 x 6.5mm thick						
Over 300mm wide, price group						
Studio A	m2	0.70	7.23	21.77	4.35	33.35
Studio B	m2	0.70	7.23	24.60	4.77	36.60
Studio C	m2	0.70	7.23	27.05	5.14	39.42

QUARRY/CERAMIC TILING

	Unit	Labour hours	Net labour (£)	Net material (£)	O'heads /profit (£)	Total (£)
Not exceeding 300mm wide, price group						
Studio A	m	0.28	2.89	6.59	1.42	10.90
Studio B	m	0.28	2.89	7.45	1.55	11.89
Studio C	m	0.28	2.89	8.20	1.66	12.75
To walls, 200 x 200 x 7mm thick						
Over 300mm wide, price group						
Studio A	m2	0.65	6.71	21.37	4.21	32.29
Studio B	m2	0.65	6.71	24.60	4.70	36.01
Not exceeding 300mm wide, price group						
Studio A	m	0.26	2.69	6.60	1.39	10.68
Studio B	m	0.26	2.69	7.45	1.52	11.66

Vitrified ceramic floor tiles to BS6431 part 6, Pilkington Dorset plain, fixing with approved adhesive, pointing with matching grout

Floors, level and to falls not exceeding 15 degrees to the horizontal, 150 x 150 x 9mm thick

	Unit	Labour hours	Net labour (£)	Net material (£)	O'heads /profit (£)	Total (£)
Over 300mm wide						
grey steel, white	m2	0.80	8.26	17.20	3.82	29.28
stone, rockface, russet blend	m2	0.80	8.26	14.33	3.39	25.98
black	m2	0.80	8.26	14.76	3.45	26.47
red	m2	0.80	8.26	12.14	3.06	23.46
Not exceeding 300mm wide						
grey steel, white	m	0.32	3.31	5.21	0.21	8.73
stone, rockface, russet blend	m	0.32	3.31	4.34	0.19	7.84
black	m	0.32	3.31	4.47	0.19	7.97
red	m	0.32	3.31	3.68	0.17	7.16

251

Ceramic tiles (cont'd)	Unit	Labour hours	Net labour (£)	Net material (£)	O'heads /profit (£)	Total (£)
Floors, to falls, crossfalls and slopes not exceeding 15 degrees to the horizontal, 150 x 150 x 9mm thick						
Over 300mm wide						
grey steel, white	m2	0.90	9.30	17.20	3.97	30.48
stone, rockface, russet blend	m2	0.90	9.30	14.33	3.54	27.17
black	m2	0.90	9.30	14.76	3.61	27.67
red	m2	0.90	9.30	12.14	3.22	24.66
not exceeding 300mm wide						
grey steel, white	m	0.36	3.72	5.21	1.34	10.27
stone, rockface, russet blend	m	0.36	3.72	4.34	1.21	9.27
black	m	0.36	3.72	4.47	1.23	9.42
red	m	0.36	3.72	3.68	1.11	8.51
Floors, exceeding 15 degrees to the horizontal						
Over 300mm wide						
grey steel, white	m2	1.00	10.33	17.20	4.13	31.66
stone, rockface, russet blend	m2	1.00	10.33	14.33	3.70	28.36
black	m2	1.00	10.33	14.76	3.76	28.85
red	m2	1.00	10.33	12.14	3.37	25.84
not exceeding 300mm wide						
grey steel, white	m	0.40	4.13	5.21	1.40	10.74
stone, rockface, russet blend	m	0.40	4.13	4.34	1.27	9.74
black	m	0.40	4.13	4.47	1.29	9.89
red	m	0.40	4.13	3.68	1.17	8.98

PLASTIC/CORK TILING/SHEETING

	Unit	Labour hours	Net labour (£)	Net material (£)	O'heads /profit (£)	Total (£)
M50 PLASTIC/CORK TILING/SHEETING						

Labour rate 10.33 per hour

Floor tiles, Marley or similar, 300 x 300mm, fixing with bitumen adhesive

Floors, level and to falls not exceeding 15 degrees from horizontal

	Unit	Labour hours	Net labour (£)	Net material (£)	O'heads /profit (£)	Total (£)
Econoflex - over 300mm wide						
series 2, 2mm thick	m2	0.24	2.48	3.10	0.84	6.42
series 2, 2.5mm thick	m2	0.26	2.69	3.55	0.94	7.18
series 4, 2mm thick	m2	0.24	2.48	3.40	0.88	6.76
series 4, 2.5mm thick	m2	0.26	2.69	3.81	0.97	7.48
Marleyflex						
2mm thick	m2	0.24	2.48	3.79	0.94	7.21
2.5mm thick	m2	0.26	2.69	4.50	1.08	8.27
Vylon						
2mm thick	m2	0.24	2.48	3.94	0.96	7.38
Sculpture						
1.8mm thick	m2	0.22	2.27	4.84	1.07	8.18
Traverline						
2.5mm thick	m2	0.26	2.69	6.29	1.35	10.33
Onyx/classic granite						
32mm thick	m2	0.24	2.48	12.32	2.22	17.02
Piazza						
2.5mm thick	m2	0.30	3.10	23.82	4.04	30.96
Medina						
3mm thick	m2	0.32	3.31	7.02	1.55	11.88

RATES FOR MEASURED WORK

Plastic tiling (cont'd)	Unit	Labour hours	Net labour (£)	Net material (£)	O'heads /profit (£)	Total (£)
Econoflex not exceeding 300mm wide						
series 2, 2mm thick	m	0.10	1.03	0.94	0.30	2.27
series 2, 2.5mm thick	m	0.11	1.14	1.08	0.33	2.55
series 4, 2mm thick	m	0.10	1.03	1.03	0.31	2.37
series 4, 2.5mm thick	m	0.11	1.14	1.15	0.34	2.63
Marleyflex						
2mm thick	m	0.10	1.03	1.15	0.33	2.51
2.5mm thick	m	0.11	1.14	1.36	0.38	2.88
Vylon						
2mm thick	m	0.10	1.03	1.19	0.33	2.55
Sculpture						
1.8mm thick	m	0.08	0.83	1.47	0.06	2.36
Traverline						
2.5mm thick	m	0.11	1.14	1.91	0.46	3.51
Onyx/classic granite						
2mm thick	m	0.10	1.03	3.72	0.71	5.46
Piazza						
2.5mm thick	m	0.11	1.14	7.22	1.25	9.61
Medina						
3mm thick	m	0.14	1.45	2.13	0.54	4.12

Increase the labour element of
all items for working in boiler
rooms or compartments not
exceeding 4m2 by 15%

VINYL TILING

**Vinyl tiles, Polyflor 300 x 300mm,
fixing with bitumen adhesive**

Floors, level and to falls not
exceeding 15 degrees from
horizontal

PLASTIC/CORK TILING/SHEETING

	Unit	Labour hours	Net labour (£)	Net material (£)	O'heads /profit (£)	Total (£)
Over 300mm wide						
Polyflor						
Premier, 2mm thick	m2	0.24	2.48	6.11	1.29	9.88
XL, 2mm thick	m2	0.24	2.48	5.70	1.23	9.41
Super XL, 2.5mm thick	m2	0.26	2.69	6.94	1.44	11.07
Super XL, 3mm thick	m2	0.28	2.89	8.21	1.66	12.76
Antistatic, 2mm thick	m2	0.24	2.48	8.31	1.62	12.41
Static conductive, 2mm thick	m2	0.24	2.48	14.69	0.43	17.60
Not exceeding 300mm wide						
Polyflor						
Premier, 2mm thick	m	0.24	2.48	1.85	0.11	4.44
XL, 2mm thick	m	0.24	2.48	1.73	0.11	4.32
Super XL, 2.5mm thick	m	0.26	2.69	2.10	0.12	4.91
Super XL, 3mm thick	m	0.28	2.89	2.49	0.13	5.51
Antistatic, 2mm thick	m	0.24	2.48	2.52	0.12	5.13
Static conductive, 2mm thick	m	0.24	2.48	4.45	0.17	7.10

Increase the labour element of all
items for working in boiler rooms
or compartments not exceeding
4m2 by 15%

VINYL SHEETING

**Marley vinyl floor coverings,
with approved adhesive, welded
joints**

Floors, level and to falls not
exceeding 15 degrees from
horizontal

over 300mm wide	Unit	Labour hours	Net labour (£)	Net material (£)	O'heads /profit (£)	Total (£)
Marleyflor, 2mm thick	m2	0.20	2.07	4.95	1.05	8.07
HD vinyl, 2mm thick	m2	0.20	2.07	6.20	1.24	9.51
HD vinyl, 2.5mm thick	m2	0.25	2.58	7.06	1.45	11.09
HD acoustic	m2	1.28	13.22	7.01	3.03	23.26
HD Hitech, 2mm thick	m2	0.20	2.07	8.31	1.56	11.94

RATES FOR MEASURED WORK

Vinyl sheeting (cont'd)	Unit	Labour hours	Net labour (£)	Net material (£)	O'heads /profit (£)	Total (£)
not exceeding 300mm wide						
Marleyflor, 2mm thick	m	0.08	0.83	1.49	0.06	2.38
HD vinyl, 2mm thick	m	0.08	0.83	1.86	0.07	2.76
HD vinyl, 2.5mm thick	m	0.10	1.03	2.12	0.08	3.23
HD acoustic	m	0.12	1.24	2.01	0.08	3.33
HD Hitech, 2mm thick	m	0.08	0.83	2.49	0.08	3.40

Increase the labour element of all
items for working in boiler rooms
or compartments not exceeding
4m2 by 15%

DECORATIVE PAPERS/FABRICS

	Unit	Labour hours	Net labour (£)	Net material (£)	O'heads /profit (£)	Total (£)

M52 DECORATIVE PAPERS/FABRICS

Labour rate 8.00 per hour

Prepare, size, apply adhesive,
supply and hang paper to plastered
walls and columns with butt joints

	Unit	Labour hours	Net labour	Net material	O'heads/profit	Total
lining paper (1.00 per roll)						
areas under 0.5m2	m2	0.29	2.32	0.24	0.38	2.94
areas over 0.5m2	m2	0.25	2.00	0.24	0.06	2.30
washable paper (2.50 per roll)						
areas under 0.5m2	m2	0.30	2.40	0.59	0.45	3.44
areas over 0.5m2	m2	0.26	2.08	0.59	0.07	2.74
vinyl paper (5.00 per roll)						
areas under 0.5m2	m2	0.30	2.40	1.18	0.54	4.12
areas over 0.5m2	m2	0.26	2.08	1.18	0.08	3.34
embossed paper (4.00 per roll)						
areas under 0.5m2	m2	0.30	2.40	0.94	0.50	3.84
areas over 0.5m2	m2	0.26	2.08	0.94	0.08	3.10
embossed paper (6.00 per roll)						
areas under 0.5m2	m2	0.30	2.40	1.41	0.57	4.38
areas over 0.5m2	m2	0.26	2.08	1.41	0.09	3.58
textured paper (4.00 per roll)						
areas under 0.5m2	m2	0.30	2.40	0.94	0.50	3.84
areas over 0.5m2	m2	0.26	2.08	0.94	0.08	3.10
textured paper (5.00 per roll)						
areas under 0.5m2	m2	0.30	2.40	1.18	0.54	4.12
areas over 0.5m2	m2	0.26	2.08	1.18	0.08	3.34
hessian paper (3.00 per m2)						
areas under 0.5m2	m2	0.55	4.40	3.30	1.16	8.86
areas over 0.5m2	m2	0.48	3.84	3.30	1.07	8.21
hessian paper (4.00 per m2)						
areas under 0.5m2	m2	0.55	4.40	4.40	1.32	10.12
areas over 0.5m2	m2	0.48	3.84	4.40	1.24	9.48

RATES FOR MEASURED WORK

Paperhanging to walls (cont'd)	Unit	Labour hours	Net labour (£)	Net material (£)	O'heads /profit (£)	Total (£)
textile wall covering (8.00 per m2)						
areas under 0.5m2	m2	0.35	2.80	8.80	1.74	13.34
areas over 0.5m2	m2	0.30	2.40	8.80	1.68	12.88
textile wall covering (10.00 per m2)						
areas under 0.5m2	m2	0.35	2.80	11.00	2.07	15.87
areas over 0.5m2	m2	0.30	2.40	11.00	2.01	15.41
suede paper (10.00 per m2)						
areas under 0.5m2	m2	0.46	3.68	11.00	2.20	16.88
areas over 0.5m2	m2	0.40	3.20	11.00	2.13	16.33
Prepare, size, apply adhesive, supply and hang paper to plastered ceilings and beams with butt joints						
lining paper (1.00 per roll)						
areas under 0.5m2	m2	0.35	2.80	0.22	0.45	3.47
areas over 0.5m2	m2	0.30	2.40	0.22	0.39	3.01
washable paper (2.50 per roll)						
areas under 0.5m2	m2	0.36	2.88	0.54	0.51	3.93
areas over 0.5m2	m2	0.31	2.48	0.54	0.08	3.10
vinyl paper (5.00 per roll)						
areas under 0.5m2	m2	0.36	2.88	1.08	0.59	4.55
areas over 0.5m2	m2	0.31	2.48	1.08	0.09	3.65
embossed paper (4.00 per roll)						
areas under 0.5m2	m2	0.36	2.88	0.86	0.56	4.30
areas over 0.5m2	m2	0.31	2.48	0.86	0.08	3.42
embossed paper (6.00 per roll)						
areas under 0.5m2	m2	0.36	2.88	1.29	0.63	4.80
areas over 0.5m2	m2	0.31	2.48	1.08	0.53	4.09
textured paper (4.00 per roll)						
areas under 0.5m2	m2	0.36	2.88	0.86	0.56	4.30
areas over 0.5m2	m2	0.31	2.48	0.86	0.08	3.42

DECORATIVE PAPERS/FABRICS

	Unit	Labour hours	Net labour (£)	Net material (£)	O'heads /profit (£)	Total (£)
textured paper (5.00 per roll)						
areas under 0.5m2	m2	0.36	2.88	1.08	0.59	4.55
areas over 0.5m2	m2	0.31	2.48	1.08	0.09	3.65

	Unit	Labour hours	Net labour (£)	Net material (£)	O'heads /profit (£)	Total (£)

M60 PAINTING

COMPOSITE RATES

Labour rate 8.00 per hour

The following composite rates are intended to help the estimator where rates are required for items involving more than one painting operation

The rates are generally based upon the information given in the previous pages, and the comments made in the early part of this chapter still apply

Two coats matt white emulsion paint, surfaces over 300mm girth

Brickwork

	Unit	Labour hours	Net labour	Net material	O'heads /profit	Total
walls	m2	0.30	2.40	0.78	0.48	3.66
walls in staircase areas	m2	0.34	2.72	0.78	0.53	4.03

Plaster

	Unit	Labour hours	Net labour	Net material	O'heads /profit	Total
walls	m2	0.22	1.76	0.64	0.36	2.76
walls in staircase areas	m2	0.26	2.08	0.64	0.41	3.13
ceilings	m2	0.30	2.40	0.64	0.46	3.50
ceilings in staircase areas	m2	0.32	2.56	0.64	0.48	3.68

Embossed paper

	Unit	Labour hours	Net labour	Net material	O'heads /profit	Total
walls	m2	0.24	1.92	0.70	0.39	3.01
walls in staircase areas	m2	0.30	2.40	0.70	0.46	3.56
ceilings	m2	0.32	2.56	0.70	0.49	3.75
ceilings in staircase areas	m2	0.36	2.88	0.70	0.54	4.12

PAINTING

	Unit	Labour hours	Net labour (£)	Net material (£)	O'heads /profit (£)	Total (£)
Two coats silk brilliant white emulsion paint, surfaces over 300mm girth						
Brickwork						
walls	m2	0.30	2.40	0.78	0.48	3.66
walls in staircase areas	m2	0.34	2.72	0.78	0.53	4.03
Plaster						
walls	m2	0.22	1.76	0.64	0.36	2.76
walls in staircase areas	m2	0.26	2.08	0.64	0.41	3.13
ceilings	m2	0.30	2.40	0.64	0.46	3.50
ceilings in staircase areas	m2	0.32	2.56	0.64	0.48	3.68
Embossed paper						
walls	m2	0.24	1.92	0.70	0.39	3.01
walls in staircase areas	m2	0.30	2.40	0.70	0.46	3.56
ceilings	m2	0.32	2.56	0.70	0.49	3.75
ceilings in staircase areas	m2	0.36	2.88	0.70	0.54	4.12
One coat primer sealer, two coats matt coloured emulsion paint, surfaces over 300mm girth						
Brickwork						
walls	m2	0.54	4.32	0.96	0.79	6.07
walls in staircase areas	m2	0.60	4.80	0.96	0.86	6.62
Plaster						
walls	m2	0.43	3.44	0.66	0.61	4.71
walls in staircase areas	m2	0.50	4.00	0.66	0.70	5.36
ceilings	m2	0.55	4.40	0.66	0.76	5.82
ceilings in staircase areas	m2	0.59	4.72	0.66	0.81	6.19

RATES FOR MEASURED WORK

	Unit	Labour hours	Net labour (£)	Net material (£)	O'heads /profit (£)	Total (£)
One coat primer sealer, one oil based undercoat, one coat white eggshell finish, surfaces over 300mm girth						
Brickwork						
walls	m2	0.60	4.80	1.33	0.92	7.05
walls in staircase areas	m2	0.64	5.12	1.33	0.97	7.42
Plaster						
walls	m2	0.53	4.24	0.98	0.78	6.00
walls in staircase areas	m2	0.58	4.64	0.98	0.84	6.46
ceilings	m2	0.61	4.88	0.98	0.88	6.74
ceilings in staircase areas	m2	0.64	5.12	0.98	0.91	7.01
One coat primer sealer, one oil based undercoat, one white gloss coat, surfaces over 300mm girth						
Brickwork						
walls	m2	0.60	4.80	1.31	0.92	7.03
walls in staircase areas	m2	0.64	5.12	1.31	0.96	7.39
Plaster						
walls	m2	0.53	4.24	0.96	0.78	5.98
walls in staircase areas	m2	0.58	4.64	0.96	0.84	6.44
ceilings	m2	0.61	4.88	0.96	0.88	6.72
ceilings in staircase areas	m2	0.64	5.12	0.96	0.91	6.99
Embossed paper						
walls	m2	0.56	4.48	1.11	0.84	6.43
walls in staircase areas	m2	0.62	4.96	1.11	0.91	6.98
ceilings	m2	0.64	5.12	1.11	0.93	7.16
ceilings in staircase areas	m2	0.69	5.52	1.11	0.99	7.62

PAINTING

	Unit	Labour hours	Net labour (£)	Net material (£)	O'heads /profit (£)	Total (£)
One coat aluminium wood primer, one oil based undercoat, one coat white eggshell finish, wood						
General surfaces						
over 300mm girth	m2	0.69	5.52	1.07	0.99	7.58
isolated surfaces not exceeding 300mm girth	m	0.26	2.08	0.36	0.37	2.81
isolated areas not exceeding 0.5m2	nr	0.36	2.88	0.54	0.51	3.93
Windows, screens, glazed doors and the like						
panes area not exceeding 0.1m2	m2	1.66	13.28	0.36	2.05	15.69
panes area 0.1 to 5m2	m2	1.43	11.44	0.54	1.80	13.78
panes area 0.5 to 1m2	m2	1.23	9.84	0.72	1.58	12.14
panes area exceeding 1m2	m2	1.01	8.08	0.80	1.33	10.21

RATES FOR MEASURED WORK

	Unit	Labour hours	Net labour (£)	Net material (£)	O'heads /profit (£)	Total (£)
One coat acrylic primer, one oil based undercoat, one coat white gloss finish, wood						
General surfaces						
over 300mm girth	m2	0.69	5.52	0.99	0.98	7.49
isolated surfaces not exceeding 300mm girth	m	0.26	2.08	0.33	0.36	2.77
isolated areas not exceeding 0.5m2	nr	0.36	2.88	0.50	0.51	3.89
Windows, screens, glazed doors and the like						
panes area not exceeding 0.1m2	m2	1.66	13.28	0.33	2.04	15.65
panes area 0.1 to 0.5m2	m2	1.43	11.44	0.50	1.79	13.73
panes area 0.5 to 1m2	m2	1.23	9.84	0.66	1.57	12.08
panes area exceeding 1m2	m2	1.01	8.08	0.72	1.32	10.12
One coat acrylic primer, one oil based undercoat, one coat coloured gloss finish, wood						
General surfaces						
over 300mm girth	m2	0.69	5.52	0.99	0.98	7.49
isolated surfaces not exceeding 300mm girth	m	0.26	2.08	0.33	0.36	2.77
isolated areas not exceeding 0.5m2	nr	0.36	2.88	0.50	0.08	3.46
Windows, screens, glazed doors and the like						
panes area not exceeding 0.1m2	m2	1.66	13.28	0.33	2.04	15.65
panes area 0.1 to 0.5m2	m2	1.43	11.44	0.50	1.79	13.73
panes area 0.5 to 1m2	m2	1.23	9.84	0.66	1.57	12.08
panes area exceeding 1m2	m2	1.01	8.08	0.72	1.32	10.12

PAINTING

	Unit	Labour hours	Net labour (£)	Net material (£)	O'heads /profit (£)	Total (£)
One coat zinc phosphate primer, one oil based undercoat, one coat white eggshell, metal						
General surfaces						
over 300mm girth	m2	0.74	5.92	0.92	1.03	7.87
isolated surfaces not exceeding 300mm girth	m	0.28	2.24	0.30	0.38	2.92
isolated areas not exceeding 0.5m2	nr	0.38	3.04	0.46	0.53	4.03
Windows, screens, glazed doors and the like						
panes area not exceeding 0.1m2	m2	1.66	13.28	0.30	2.04	15.62
panes area 0.1 to 0.5m2	m2	1.43	11.44	0.43	1.78	13.65
panes area 0.5 to 1m2	m2	1.23	9.84	0.60	1.57	12.01
panes area exceeding 1m2	m2	1.06	8.48	0.67	1.37	10.52
Structural metalwork, general surfaces						
over 300mm girth	m2	0.84	6.72	0.92	1.15	8.79
isolated surfaces not exceeding 300mm girth	m	0.31	2.48	0.30	0.42	3.20
isolated areas not exceeding 0.5m2	nr	0.41	3.28	0.46	0.09	3.83
Structural metalwork, members of roof trusses, lattice girders, purlins and the like						
over 300mm girth	m2	0.97	7.76	0.92	1.30	9.98
isolated areas not exceeding 300mm girth	m	0.35	2.80	0.30	0.08	3.18
isolated areas not exceeding 0.5m2	nr	0.39	3.12	0.46	0.54	4.12

RATES FOR MEASURED WORK

	Unit	Labour hours	Net labour (£)	Net material (£)	O'heads /profit (£)	Total (£)
Radiators, panel type						
over 300mm, girth	m2	0.98	7.84	0.92	0.22	8.98
isolated surfaces not exceeding 300mm girth	m	0.31	2.48	0.30	0.42	3.20
isolated areas not exceeding 0.5m2	nr	0.44	3.52	0.46	0.60	4.58
Radiators, column type						
over 300mm girth	m2	1.03	8.24	0.92	1.37	10.53
isolated surfaces not exceeding 300mm girth	m	0.38	3.04	0.30	0.08	3.42
isolated areas not exceeding 0.5m2	nr	0.52	4.16	0.46	0.69	5.31

One coat quick drying metal primer, one oil based undercoat, one coat brilliant white gloss finish, metal

General surfaces

	Unit	Labour hours	Net labour (£)	Net material (£)	O'heads /profit (£)	Total (£)
over 300mm girth	m2	0.78	6.24	0.90	1.07	8.21
isolated surfaces not exceeding 300mm girth	m	0.28	2.24	0.30	0.38	2.92
isolated areas not exceeding 0.5m2	nr	0.38	3.04	0.45	0.52	4.01

Windows, screens, glazed doors and the like

	Unit	Labour hours	Net labour (£)	Net material (£)	O'heads /profit (£)	Total (£)
panes area not exceeding 0.1m2	m2	1.66	13.28	0.30	0.34	13.92
panes area 0.1 to 0.5m2	m2	1.43	11.44	0.45	1.78	13.67
panes area 0.5 to 1m2	m2	1.23	9.84	0.60	1.57	12.01
panes area exceeding 1m2	m2	1.06	8.48	0.67	1.37	10.52

PAINTING

	Unit	Labour hours	Net labour (£)	Net material (£)	O'heads /profit (£)	Total (£)
Structural metalwork, general surfaces						
over 300mm girth	m2	0.84	6.72	0.90	1.14	8.76
isolated surfaces not exceeding 300mm girth	m	0.31	2.48	0.30	0.42	3.20
isolated areas not exceeding 0.5m2	nr	0.41	3.28	0.45	0.09	3.82
Structural metalwork, members of roof trusses, lattice girders, purlins and the like						
over 300mm girth	m2	0.97	7.76	0.90	1.30	9.96
isolated surfaces not exceeding 300mm girth	m	0.35	2.80	0.30	0.46	3.57
isolated areas not exceeding 0.5m2	nr	0.39	3.12	0.45	0.54	4.11
Radiators, panel type						
over 300mm girth	m2	0.98	7.84	0.90	1.31	10.05
isolated surfaces not exceeding 300mm girth	m	0.31	2.48	0.30	0.42	3.20
isolated areas not exceeding 0.5m2	nr	0.44	3.52	0.45	0.60	4.57
Radiators, column type						
over 300mm girth	m2	1.03	8.24	0.90	1.37	10.51
isolated surfaces not exceeding 300mm girth	m	0.38	3.04	0.30	0.50	3.84
isolated areas not exceeding 0.5m2	nr	0.52	4.16	0.45	0.69	5.30

RATES FOR MEASURED WORK

	Unit	Labour hours	Net labour (£)	Net material (£)	O'heads /profit (£)	Total (£)
One coat calcium plumbate primer, one oil based undercoat, one coat brilliant white eggshell finish, metal						
General surfaces						
over 300mm girth	m2	0.70	5.60	1.26	1.03	7.89
isolated surfaces not exceeding 300mm girth	m	0.28	2.24	0.42	0.40	3.06
isolated areas not exceeding 0.5m2	nr	0.36	2.88	0.63	0.53	4.04
Windows, screens, glazed doors and the like						
panes area not exceeding 0.1m2	m2	1.60	12.80	0.42	1.98	15.20
panes area 0.1 to 0.5m2	m2	1.39	11.12	0.63	1.76	13.51
panes area 0.5 to 1m2	m2	1.17	9.36	0.84	0.26	10.46
panes area exceeding 1m2	m2	1.00	8.00	0.94	1.34	10.28
Structural metalwork, general surfaces						
over 300mm girth	m2	0.75	6.00	1.26	1.09	8.35
isolated surfaces not exceeding 300mm girth	m	0.29	2.32	0.42	0.41	3.15
isolated areas not exceeding 0.5m2	nr	0.37	2.96	0.63	0.54	4.13
Structural metalwork, members of roof trusses, lattice girders, purlins and the like						
over 300mm girth	m2	1.03	8.24	1.26	1.43	10.93
isolated surfaces not exceeding 300mm girth	m	0.35	2.80	0.42	0.48	3.70
isolated areas not exceeding 0.5m2	nr	0.47	3.76	0.63	0.66	5.05

PAINTING

	Unit	Labour hours	Net labour (£)	Net material (£)	O'heads /profit (£)	Total (£)
Radiators, panel type						
over 300mm girth	m2	0. 84	6. 72	1. 26	1. 20	9. 18
isolated surfaces not exceeding 300mm girth	m	0. 31	2. 48	0. 42	0. 43	3. 34
isolated areas not exceeding 0.5m2	nr	0. 42	3. 36	0. 63	0. 60	4. 59
Radiators, column type						
over 300mm girth	m2	0. 98	7. 84	1. 26	1. 36	10. 47
isolated surfaces not exceeding 300mm girth	m	0. 38	3. 04	0. 42	0. 52	3. 98
isolated areas not exceeding 0.5m2	nr	0. 50	4. 00	0. 63	0. 69	5. 32

269

N

FURNITURE/EQUIPMENT

MATERIAL COSTS

	Unit	Material supply (£)	Material waste (%)	Total (£)
MATERIAL COSTS				
Baths				
Acrylic				
3mm thick, white and coloured	nr	88.19	2.5	90.39
5mm thick, white and coloured	nr	116.39	2.5	119.30
Pressed steel, vitreous enamelled, standard gauge				
white	nr	91.20	2.5	93.48
coloured	nr	100.68	2.5	103.20
Pressed steel vitreous enamelled heavy gauge				
white	nr	242.37	2.5	248.43
coloured	nr	278.78	2.5	285.75
Bath panel, hardboard				
end	nr	2.51	2.5	2.57
side	nr	5.02	2.5	5.15
Bath panel, polystyrene				
end	nr	14.63	2.5	15.00
side	nr	22.30	2.5	22.86
Aluminium angle strip	m	1.65	2.5	1.69
Basins and pedestals				
560 x 430mm				
wall-mounted	nr	32.24	2.5	33.05
with pedestal	nr	51.52	2.5	52.81
590 x 440mm, white medical wall-mounted	nr	61.91	2.5	63.46

RATES FOR MEASURED WORK

	Unit	Material supply (£)	Material waste (%)	Total (£)
495 x 420mm corner wall-mounted white or coloured	nr	32.47	2.5	33.28
550 x 450mm, vanity basin white or coloured	nr	57.24	2.5	58.67

Sinks

Stainless steel sink unit

	Unit			
single bowl, double drainer, 1000 x 500mm	nr	84.00	2.5	86.10
single bowl, double drainer 1500 x 500mm	nr	84.00	2.5	86.10
double bowl, single drainer 1500x 500mm	nr	247.84	2.5	254.04

Fireclay sink, Belfast pattern

455 x 380 x 205mm	nr	81.28	2.5	83.31
610 x 455 x 255mm	nr	129.13	2.5	132.36
760 x 455 x 255mm	nr	167.43	2.5	171.62

Fireclay sink, shelf pattern

610 x 535 x 255mm	nr	149.29	2.5	153.02

WC suites

Close-coupled, white

washdown	nr	148.33	2.5	152.04
siphonic	nr	269.84	2.5	276.59

Close-coupled, coloured

washdown	nr	164.79	2.5	168.91
siphonic	nr	274.05	2.5	280.90

Bidets

Freestanding, plain rim, white	nr	107.81	2.5	110.51
Freestanding, plain rim, coloured	nr	138.97	2.5	142.44

MATERIAL COSTS

	Unit	Material supply (£)	Material waste (%)	Total (£)
Taps				
Aztec, chromium-plated				
13mm	pr	31. 19	2. 5	31. 97
19mm	pr	44. 73	2. 5	45. 85
Amazon, chromium-plated				
13mm	pr	11. 10	2. 5	11. 38
19mm	pr	12. 30	2. 5	12. 61
Sink pillar taps				
Peglers	pr	19. 95	2. 5	20. 45
Streamline	pr	32. 81	2. 5	33. 63
Mixer taps				
Basin, Streamline	pr	38. 85	2. 5	39. 82
Basin, Swanlux	pr	19. 16	2. 5	19. 64
Bath with handspray	pr	123. 90	2. 5	127. 00
Bath, filler only	pr	75. 60	2. 5	77. 49
Bath with handspray	pr	119. 70	2. 5	122. 69
Sink with spout	pr	35. 18	2. 5	36. 06
Sink with spout	pr	69. 30	2. 5	71. 03
Sink with monoblock	pr	69. 83	2. 5	71. 58
Sink with hot rinse	pr	69. 83	2. 5	71. 58
Bidet fittings				
Monoblock and spray	nr	115. 50	2. 5	118. 39
Waste disposal units				
Tweeny	nr	184. 80	2. 5	189. 42
Carror	nr	108. 15	2. 5	110. 85

RATES FOR MEASURED WORK

	Unit	Material supply (£)	Material waste (%)	Total (£)
Valves				
Mira valves				
type RADA 15	nr	360.91	2.5	369.93
type RADA 17	nr	428.59	2.5	439.30
type RADA 20	nr	473.23	2.5	485.06
type RADA 25	nr	654.64	2.5	671.01
type 88	nr	84.00	0.0	84.00
type 88B	nr	84.00	0.0	84.00
Shower kits				
Mira				
type SF1/100	nr	52.50	2.5	53.81
type NW SF1/300	nr	52.50	2.5	53.81
type 881	nr	357.00	2.5	365.93
Excel	nr	451.50	2.5	462.79
Heatrae shower systems				
Cameo plus	nr	73.92	1.0	74.66
Carousel standard	nr	121.28	1.0	122.49
Accolade electronic standard	nr	189.99	1.0	191.89
Shower trays				
Fireclay	nr	129.15	1.0	130.44
Acrylic	nr	48.30	1.0	48.78
'Daryl' shower enclosure				
pivot door, gold	nr	473.55	1.0	478.29
pivot door, silver	nr	252.00	1.0	254.52
side panel, gold	nr	182.70	1.0	184.53
'Matki' bathscreen				
NWMV 700, white	nr	218.40	1.0	220.58

MATERIAL COSTS

	Unit	Material supply (£)	Material waste (%)	Total (£)
'Matki' shower enclosures				
XE75, corner entry, silver	nr	626. 85	1. 0	633. 12
NW7375, side panel, silver	nr	245. 70	1. 0	248. 16
NW7375, side panel, gold	nr	262. 50	1. 0	265. 13
CH7578, side panel	nr	161. 15	1. 0	162. 76

RATES FOR MEASURED WORK

	Unit	Labour hours	Net labour (£)	Net material (£)	O'heads /profit (£)	Total (£)
N11 DOMESTIC KITCHEN FITTINGS						
Labour rate 8.00 per hour						
Kitchen fittings						
Fix only kitchen fittings						
base units 600mm deep						
600mm long	nr	0.50	4.00	0.00	0.60	4.60
1000mm long	nr	0.60	4.80	0.00	0.72	5.52
1200mm long	nr	0.70	5.60	0.00	0.84	6.44
sink base units 600mm deep						
600mm long	nr	0.60	4.80	0.00	0.72	5.52
1000mm long	nr	0.70	5.60	0.00	0.84	6.44
1200mm long	nr	0.80	6.40	0.00	0.96	7.36
tall storage units 2100mm high, 600mm deep						
600mm long	nr	0.90	7.20	0.00	1.08	8.28
walls units 300mm deep						
600mm long	nr	1.20	9.60	0.00	1.44	11.04
1000mm long	nr	1.30	10.40	0.00	1.56	11.96
1200mm long	nr	1.40	11.20	0.00	1.68	12.88

	Unit	Labour hours	Net labour (£)	Net material (£)	O'heads /profit (£)	Total (£)

N13 SANITARY APPLIANCES

Labour rate 9.60 per hour

BATHS

Acrylic, reinforced, complete
with 2nr chromium-plated clips,
waste, overflow with chain and
plastic plug, excluding taps and
trap

1700mm long
white	nr	2.55	24.48	92.66	17.57	134.71
coloured	nr	2.55	24.48	122.28	22.01	168.77

Standard gauge (1.6mm) pressed
steel vitreous enamelled, complete
with 2nr chromium-plated grips,
waste overflow with chain and
plastic plug, excluding taps and
trap

1700mm long
white	nr	2.65	25.44	95.82	18.19	139.45
coloured	nr	2.65	25.44	105.79	19.68	150.91

Heavy gauge (2.3mm) pressed steel
vitreous enamelled to BS1344 (no
grips) slip-resistant, waste
overflow with chain and plastic
plug, excluding taps and trap

1700mm long
white	nr	3.15	30.24	254.64	42.73	327.61
coloured	nr	3.15	30.24	292.89	48.47	371.60

RATES FOR MEASURED WORK

	Unit	Labour hours	Net labour (£)	Net material (£)	O'heads /profit (£)	Total (£)
Bath accessories						
Bath panels, enamelled hardboard, fixing with chromium dome-headed screws, trimming as required						
end panel, BP 2.06	nr	0.30	2.88	2.63	0.83	6.34
side panel, BP 3.92	nr	0.45	4.32	5.28	1.44	11.04
Bath panels, moulded high-impact polystyrene, fixing in position, for trimming as required						
end panel	nr	0.25	2.40	15.36	2.66	20.42
side panel	nr	0.40	3.84	23.43	4.09	31.36
Angle strip polished aluminium, fixing with chromium-plated dome-headed screws, cutting to length						
25 x 25 x 560mm long	nr	0.30	2.88	1.73	0.69	5.30
BASINS AND PEDESTALS						
Wash basin, vitreous china, complete with chromium-plated waste, overflow with chain and plastic plug, excluding taps and trap						
560 x 430mm, white or coloured						
wall-mounted on brackets	nr	1.75	16.80	33.88	7.60	58.28
pedestal-mounted, bedded in mastic, screwed to wall	nr	2.05	19.68	54.13	11.07	84.88
590 x 475mm, white or coloured						
wall-mounted on brackets	nr	1.80	17.28	65.05	12.35	94.68
pedestal-mounted, bedded in mastic, screwed to wall	nr	2.10	20.16	34.11	8.14	62.41

SANITARY APPLIANCES

	Unit	Labour hours	Net labour (£)	Net material (£)	O'heads /profit (£)	Total (£)
590 x 440mm, white medical wall-mounted on brackets	nr	1.85	17.76	65.05	12.42	95.23

SINKS

Stainless steel sink unit sinks, complete with inset chromium-plated waste, overflow with chain and plastic plug, excluding taps and trap, fixing to sink top with proprietary clips

	Unit	Labour hours	Net labour (£)	Net material (£)	O'heads /profit (£)	Total (£)
single bowl with single drainer 1000 x 500mm	nr	1.80	17.28	88.25	15.83	121.36
single bowl with double drainer 1500 x 500mm	nr	1.90	18.24	88.25	15.97	122.46
double bowl with single drainer 1500 x 500mm	nr	1.95	18.72	260.41	41.87	321.00

Fireclay sinks, BS1206, white glazed, complete with chromium-plated waste, chain and plastic plug, excluding taps and trap, wall-mounted on brackets

	Unit	Labour hours	Net labour (£)	Net material (£)	O'heads /profit (£)	Total (£)
Belfast pattern						
455 x 380 x 205mm	nr	2.25	21.60	85.39	16.05	123.04
610 x 455 x 255mm	nr	2.35	22.56	135.67	23.73	181.96
760 x 455 x 255mm	nr	2.70	25.92	175.91	30.27	232.10
Shelf pattern						
610 x 535 x 255mm	nr	2.50	24.00	156.85	27.13	207.98

RATES FOR MEASURED WORK

	Unit	Labour hours	Net labour (£)	Net material (£)	O'heads /profit (£)	Total (£)
WC SUITES						
Close-coupled, vitreous china with plastic seat, comprising 9 litre cistern, ball valve, flush pipe, pan and all brackets, connection to drain						
white						
washdown type	nr	2.35	22.56	155.83	26.76	205.15
siphonic type	nr	2.35	22.56	283.50	45.91	351.97
coloured						
washdown type	nr	2.35	22.56	173.13	29.35	225.04
siphonic type	nr	2.35	22.56	287.92	46.57	357.05
BIDETS						
Freestanding plain rim, vitreous china, excluding fittings						
white	nr	1.95	18.72	113.26	19.80	151.78
coloured	nr	1.95	18.72	146.00	24.71	189.43
TAPS						
Basin and bath pillar taps						
Aztec chromium-plated						
13mm	nr	0.35	3.36	32.77	5.42	41.55
19mm	nr	0.35	3.36	47.00	7.55	57.91
Amazon chromium-plated						
13mm	nr	0.35	3.36	11.66	2.25	17.27
19mm	nr	0.35	3.36	12.93	2.44	18.73
Sink pillar taps						
Pegler's chromium-plated	nr	0.35	3.36	20.96	0.61	24.93
Streamline chromium-plated	nr	0.35	3.36	34.51	5.68	43.55

SANITARY APPLIANCES

	Unit	Labour hours	Net labour (£)	Net material (£)	O'heads /profit (£)	Total (£)
Basin mixer taps						
Streamline chromium-plated, 3 hole pop-up waste with fixed head	nr	0.50	4.80	40.82	6.84	52.46
Swanlux chromium-plated, monoblock pop-up waste with fixed head	nr	0.50	4.80	46.04	7.63	58.47
Bath mixer taps						
Streamline chromium-plated, deck pattern with hose and handspray	nr	0.70	6.72	41.57	7.24	55.53
Streamline chromium-plated, deck pattern, filler only	nr	0.55	5.28	48.81	8.11	62.20
Streamline chromium-plated, deck pattern with handspray	nr	0.70	6.72	64.96	10.75	82.43
Sink mixer taps						
Streamline chromium-plated, deck pattern with swivel spout	nr	0.40	3.84	30.24	5.11	39.19
Streamline chromium-plated, deck pattern with swivel spout	nr	0.40	3.84	39.44	6.49	49.77
Streamline studio chromium-plated, monoblock with swivel spout	nr	0.40	3.84	48.79	7.89	60.52
Streamline studio chromium-plated, monoblock with swivel spout, hot rinse attachment	nr	0.50	4.80	60.63	9.81	75.24
Bidet fittings						
Twyford's Aztec chromium-plated, monoblock and spray	nr	0.65	6.24	121.35	3.19	130.78

RATES FOR MEASURED WORK

	Unit	Labour hours	Net labour (£)	Net material (£)	O'heads /profit (£)	Total (£)
WASTE DISPOSAL UNITS						
For metal sinks with 89mm waste outlet, excluding electrical connections						
Tweeny	nr	2.00	19.20	194.16	32.00	245.36
Carror	nr	1.50	14.40	113.62	19.20	147.22
VALVES						
Mira, thermostatic mixing valves, fixing to wall						
type RADA15	nr	0.50	4.80	379.18	57.60	441.58
type RADA17	nr	0.50	4.80	450.28	68.26	523.34
type RADA 20	nr	0.50	4.80	466.44	70.69	541.93
type RADA 25	nr	0.50	4.80	687.79	103.89	796.48
type 88	nr	0.50	4.80	88.36	13.97	107.13
type 88B	nr	0.50	4.80	88.36	13.97	107.13
SHOWER KITS						
Mira, fixing to wall						
type SF1/100 handset with flexible tube, connector and holder	nr	0.50	4.80	55.16	8.99	68.95
type SF1/300 flexible tube with adjustable handspray, sliding bar, NWSF3	nr	0.75	7.20	55.16	9.35	71.71
type Powerpak 881, power shower, with mixer, pump, handset and gel dispenser	nr	1.00	9.60	375.08	57.70	442.38
type Excel, power shower, with thermostatic mixer pump, handset and gel dispenser	nr	1.10	10.56	474.36	72.74	557.66

SANITARY APPLIANCES

	Unit	Labour hours	Net labour (£)	Net material (£)	O'heads /profit (£)	Total (£)
ELECTRICAL SHOWER SYSTEMS						
Heatrae Sadia, fixing to wall, excluding electrical connections F7						
type Cameo-7kW	nr	0.75	7.20	76.12	12.50	95.82
type Carousel-7kW	nr	0.75	7.20	125.55	19.91	152.66
type Accolade-7kW	nr	0.75	7.20	196.69	30.58	234.47
SHOWERS						
Trays						
white glazed fireclay, fixing in postion, 760 x 760 x 180mm, white	nr	2.50	24.00	133.69	3.94	161.63
acrylic, fixing in position, 760 x 760 x 180mm	nr	1.50	14.40	50.01	1.61	66.02
Enclosures						
'Daryl' shower enclosure, fixing ends to walls, safety glass						
pivot door, silver	nr	1.00	9.60	490.24	74.98	574.82
side panel, gold	nr	0.70	6.72	260.88	40.14	307.74
side panel, silver	nr	0.70	6.72	189.14	29.38	225.24
'Matki bathscreen', fixing ends to walls						
NWMV700, white	nr	1.00	9.60	226.10	35.35	271.05
'Matki' shower enclosures, fixing ends to walls, toughened glass (unless otherwise stated)						
XE75, corner entry, silver	nr	1.00	9.60	648.95	98.78	757.33
MW7375, side panel, silver	nr	1.00	9.60	254.36	39.59	303.55
MW7375, side panel, gold	nr	0.70	6.72	271.75	41.77	320.24
CH7578, side panel	nr	0.70	6.72	166.83	26.03	199.58

P

BUILDING FABRIC SUNDRIES

MATERIAL COSTS

	Unit	Material supply (£)	Material waste (%)	Total (£)
MATERIAL COSTS				
Skirting, architraves				
19 x 50mm	m	0.56	5.0	0.59
19 x 63mm	m	0.69	5.0	0.72
25 x 50mm	m	0.97	5.0	1.02
25 x 63mm	m	1.09	5.0	1.14
25 x 75mm	m	1.15	5.0	1.21
25 x 100mm	m	1.39	5.0	1.46
25 x 125mm	m	1.61	5.0	1.69
25 x 150mm	m	1.94	5.0	2.04
Rails, moulded				
19 x 50mm	m	0.71	5.0	0.75
19 x 75mm	m	0.76	5.0	0.80
19 x 100mm	m	0.77	5.0	0.81
25 x 50mm	m	0.84	5.0	0.88
25 x 75mm	m	1.07	5.0	1.12
25 x 100mm	m	1.29	5.0	1.35
Handrail, mopstick				
50 x 50mm	m	1.70	5.0	1.79
Glazing beads				
13 x 25mm	m	0.70	5.0	0.74
19 x 36mm	m	0.87	5.0	0.91
19 x 50mm	m	0.97	5.0	1.02
Shelving worktops 19mm thick, width				
150mm	m	2.85	5.0	2.99
225mm	m	3.48	5.0	3.65
300mm	m	2.40	5.0	2.52
Shelving bearers				
19 x 50mm	m	0.65	5.0	0.68
25 x 50mm	m	0.70	5.0	0.74

RATES FOR MEASURED WORK

	Unit	Material supply (£)	Material waste (%)	Total (£)
Wrought hardwood				
Architraves, skirtings, chamfered				
19 x 50mm	m	1.65	5.0	1.73
19 x 63mm	m	2.08	5.0	2.18
25 x 50mm	m	2.91	5.0	3.06
25 x 63mm	m	3.28	5.0	3.44
25 x 75mm	m	2.44	5.0	2.56
25 x 100mm	m	4.17	5.0	4.38
25 x 125mm	m	4.83	5.0	5.07
25 x 150mm	m	5.82	5.0	6.11
Rails, moulded				
19 x 50mm	m	2.15	5.0	2.26
19 x 75mm	m	2.28	5.0	2.39
19 x 100mm	m	2.32	5.0	2.44
25 x 50mm	m	2.51	5.0	2.64
25 x 75mm	m	3.21	5.0	3.37
25 x 100mm	m	3.88	5.0	4.07
Handrail, mopstick				
50 x 50mm	m	5.09	5.0	5.34
Glazing beads				
13 x 25mm	m	2.16	5.0	2.27
19 x 36mm	m	2.61	5.0	2.74
19 x 50mm	m	2.91	5.0	3.06
Shelving worktops 19mm thick, width				
150mm	m	8.54	5.0	8.97
225mm	m	10.42	5.0	10.94
300mm	m	22.20	5.0	23.31
Shelving bearers				
19 x 50mm	m	1.95	5.0	2.05
25 x 50mm	m	2.12	5.0	2.23

MATERIAL COSTS

	Unit	Material supply (£)	Material waste (%)	Total (£)
Hardboard 3.2mm thick	m2	1.56	5.0	1.64
Hardboard 6.4mm thick	m2	3.00	5.0	3.15
Teak faced one side blockboard				
18mm thick	m2	20.04	5.0	21.04
Chipboard 12mm thick	m2	3.28	5.0	3.44
Chipboard 15mm thick	m2	4.77	5.0	5.01
Veneered one face				
plywood 4mm thick	m2	6.58	5.0	6.91
plywood 6mm thick	m2	9.02	5.0	9.47
plywood 9mm thick	m2	10.93	5.0	11.48
plywood 12mm thick	m2	13.27	5.0	13.93
Melamine faced chipboard				
15mm thick	m2	6.67	5.0	7.00
Insulation board				
12.7mm	m2	5.92	5.0	6.22
19mm	m2	7.84	5.0	8.23
25mm	m2	10.78	5.0	11.32
Fibreglass insulation				
50mm thick	m2	1.80	5.0	1.89
100mm thick	m2	1.95	5.0	2.05
Polystyrene sheeting				
12mm thick	m2	0.70	5.0	0.74
25mm thick	m2	1.45	5.0	1.52
50mm thick	m2	2.88	5.0	3.02

RATES FOR MEASURED WORK

	Unit	Labour hours	Net labour (£)	Net material (£)	O'heads /profit (£)	Total (£)
P10 SUNDRY INSULATION						
Labour rate 8.00 per hour						
Glass fibre quilt laid over ceiling joists, thickness						
50mm	m2	0.15	1.20	1.80	0.45	3.45
100mm	m2	0.16	1.28	1.95	0.48	3.71
Glass fibre quilt fixed vertically to softwood, thickness						
50mm	m2	0.18	1.44	1.80	0.49	3.73
100mm	m2	0.19	1.52	1.95	0.52	3.99
Expanded polystyrene						
Expanded polystyrene board fixed with adhesive to walls, thickness						
12mm	m2	0.60	4.80	0.70	0.82	6.32
25mm	m2	0.65	5.20	1.45	1.00	7.65
50mm	m2	0.70	5.60	2.88	1.27	9.75

	Unit	Labour hours	Net labour (£)	Net material (£)	O'heads /profit (£)	Total (£)

P20 UNFRAMED ISOLATED SUNDRY TRIMS SKIRTINGS/SUNDRY ITEMS

Labour rate 8.00 per hour

Wrought softwood

Architraves, skirtings chamfered

	Unit	Labour hours	Net labour (£)	Net material (£)	O'heads /profit (£)	Total (£)
19 x 50mm	m	0.14	1.12	0.53	0.25	1.90
19 x 63mm	m	0.14	1.12	0.69	0.27	2.08
25 x 50mm	m	0.15	1.20	0.97	0.33	2.50
25 x 63mm	m	0.15	1.20	1.09	0.34	2.63
25 x 75mm	m	0.15	1.20	1.15	0.35	2.70
25 x 100mm	m	0.17	1.36	1.39	0.41	3.16
25 x 125mm	m	0.18	1.44	1.61	0.46	3.51
25 x 150mm	m	0.18	1.44	1.94	0.51	3.89

Rails, moulded

	Unit	Labour hours	Net labour (£)	Net material (£)	O'heads /profit (£)	Total (£)
19 x 50mm	m	0.14	1.12	0.71	0.27	2.10
19 x 75mm	m	0.14	1.12	0.76	0.28	2.16
19 x 100mm	m	0.15	1.20	0.77	0.30	2.27
25 x 50mm	m	0.15	1.20	0.84	0.31	2.35
25 x 75mm	m	0.15	1.20	1.07	0.34	2.61
25 x 100mm	m	0.17	1.36	1.29	0.40	3.05

Handrail, mopstick

	Unit	Labour hours	Net labour (£)	Net material (£)	O'heads /profit (£)	Total (£)
50 x 50mm	m	0.15	1.20	1.70	0.43	3.33

Glazing beads

	Unit	Labour hours	Net labour (£)	Net material (£)	O'heads /profit (£)	Total (£)
13 x 25mm	m	0.10	0.80	0.70	0.22	1.72
19 x 36mm	m	0.10	0.80	0.87	0.25	1.92
19 x 50mm	m	0.10	0.80	0.97	0.27	2.04

Shelving worktops 19mm thick, width

	Unit	Labour hours	Net labour (£)	Net material (£)	O'heads /profit (£)	Total (£)
150mm	m	0.33	2.64	2.85	0.82	6.31
225mm	m	0.40	3.20	3.48	1.00	7.68

RATES FOR MEASURED WORK

Softwood (cont'd)	Unit	Labour hours	Net labour (£)	Net material (£)	O'heads /profit (£)	Total (£)
Shelving bearers						
19 x 50mm	m	0.15	1.20	0.65	0.28	2.13
25 x 50mm	m	0.16	1.28	0.70	0.30	2.28
Wrought hardwood						
Architraves, skirtings chamfered						
19 x 50mm	m	0.21	1.68	1.65	0.50	3.83
19 x 63mm	m	0.21	1.68	2.08	0.56	4.32
25 x 50mm	m	0.22	1.76	2.91	0.70	5.37
25 x 63mm	m	0.22	1.76	3.28	0.76	5.80
25 x 75mm	m	0.22	1.76	3.44	0.78	5.98
25 x 100mm	m	0.25	2.00	4.17	0.93	7.10
25 x 125mm	m	0.26	2.08	4.83	1.04	7.95
25 x 150mm	m	0.26	2.08	5.82	1.19	9.09
Rails, moulded						
19 x 50mm	m	0.21	1.68	2.15	0.57	4.40
19 x 75mm	m	0.21	1.68	2.28	0.59	4.55
19 x 100mm	m	0.22	1.76	2.62	0.66	5.04
25 x 50mm	m	0.22	1.76	2.51	0.64	4.91
25 x 75mm	m	0.22	1.76	3.21	0.75	5.72
25 x 100mm	m	0.25	2.00	3.88	0.88	6.76
Handrail, mopstick						
50 x 50mm	m	0.25	2.00	5.09	1.06	8.15
Glazing beads						
13 x 25mm	m	0.15	1.20	2.12	0.50	3.82
19 x 36mm	m	0.15	1.20	2.61	0.57	4.38
19 x 50mm	m	0.15	1.20	2.91	0.62	4.73
Shelving worktops 19mm thick, width						
150mm	m	0.47	3.76	8.54	1.84	14.14
225mm	m	0.55	4.40	10.42	2.22	17.04
300mm	m	0.60	4.80	22.20	4.05	31.05

294

UNFRAMED ISOLATED TRIMS/SKIRTINGS

	Unit	Labour hours	Net labour (£)	Net material (£)	O'heads /profit (£)	Total (£)
Shelving bearers						
19 x 50mm	m	0.27	2.16	1.95	0.62	4.73
25 x 50mm	m	0.25	2.00	2.12	0.62	4.74
Sheet linings and casings						
Linings or casings over 300mm wide						
hardboard 3.2mm thick	m2	0.53	4.24	1.56	0.87	6.67
hardboard 6.4mm thick	m2	0.55	4.40	3.00	1.11	8.51
teak faced blockboard 18mm thick	m2	0.60	4.80	20.04	3.73	28.57
chipboard 12mm thick	m2	0.50	4.00	3.28	1.09	8.37
chipboard 15mm thick	m2	0.61	4.88	4.77	1.45	11.10
plywood 4mm thick	m2	0.34	2.72	6.58	1.40	10.70
plywood 6mm thick	m2	0.36	2.88	9.02	1.78	13.68
plywood 9mm thick	m2	0.40	3.20	10.93	2.12	16.25
plywood 12mm thick	m2	0.46	3.68	13.27	2.54	19.49
melamine faced chipboard 15mm thick	m2	0.58	4.64	6.67	1.70	13.01
insulation board 12.7mm thick	m2	0.36	2.88	5.92	1.32	10.12
insulation board 19mm thick	m2	0.38	3.04	7.84	1.63	12.51
insulation board 25mm thick	m2	0.40	3.20	10.76	2.09	16.05
Linings or casings not exceeding 100mm wide						
hardboard 3.2mm thick	m	0.18	1.44	0.16	0.24	1.84
hardboard 6.4mm thick	m	0.18	1.44	0.30	0.26	2.00
teak faced blockboard 18mm thick	m	0.20	1.60	2.00	0.54	4.14
chipboard 12mm thick	m	0.16	1.28	0.33	0.24	1.85
chipboard 15mm thick	m	0.18	1.44	0.48	0.29	2.21
plywood 4mm thick	m	0.12	0.96	0.66	0.24	1.86
plywood 6mm thick	m	0.13	1.04	0.90	0.29	2.23
plywood 9mm thick	m	0.14	1.12	1.09	0.33	2.54
plywood 12mm thick	m	0.16	1.28	1.33	0.39	3.00
melamine faced chipboard 15mm thick	m	0.21	1.68	0.67	0.35	2.70
insulation board 12.7mm thick	m	0.14	1.12	0.59	0.26	1.97
insulation board 19mm thick	m	0.15	1.20	0.78	0.30	2.28
insulation board 25mm thick	m	0.18	1.44	1.08	0.38	2.90

RATES FOR MEASURED WORK

Hardwood (cont'd)	Unit	Labour hours	Net labour (£)	Net material (£)	O'heads /profit (£)	Total (£)
Linings or casings 100 to 200mm wide						
hardboard 3.2mm thick	m	0.21	1.68	0.32	0.30	2.30
hardboard 6.4mm thick	m	0.21	1.68	0.60	0.34	2.62
teak faced blockboard 18mm thick	m	0.24	1.92	4.00	0.89	6.81
chipboard 12mm thick	m	0.19	1.52	0.66	0.33	2.51
chipboard 15mm thick	m	0.21	1.68	0.96	0.40	3.04
plywood 4mm thick	m	0.15	1.20	1.22	0.36	2.78
plywood 6mm thick	m	0.16	1.28	1.80	0.46	3.54
plywood 9mm thick	m	0.17	1.36	2.18	0.53	4.07
plywood 12mm thick	m	0.19	1.52	2.66	0.63	4.81
melamine faced chipboard 15mm thick	m	0.24	1.92	1.34	0.49	3.75
insulation board 12.7mm thick	m	0.17	1.36	1.18	0.38	2.92
insulation board 19mm thick	m	0.18	1.44	1.56	0.45	3.45
insulation board 25mm thick	m	0.20	1.60	2.16	0.56	4.32
Linings or casings 200 to 300mm wide						
hardboard 3.2mm thick	m	0.25	2.00	0.48	0.37	2.85
hardboard 6.4mm thick	m	0.25	2.00	0.90	0.43	3.33
teak faced blockboard 18mm thick	m	0.30	2.40	6.00	1.26	9.66
chipboard 12mm thick	m	0.24	1.92	0.99	0.44	3.35
chipboard 15mm thick	m	0.25	2.00	1.44	0.52	3.96
plywood 4mm thick	m	0.20	1.60	1.98	0.54	4.12
plywood 6mm thick	m	0.21	1.68	2.70	0.66	5.04
plywood 9mm thick	m	0.22	1.76	3.27	0.75	5.78
plywood 12mm thick	m	0.24	1.92	3.99	0.89	6.80
melamine faced chipboard 15mm thick	m	0.28	2.24	1.41	0.55	4.20
insulation board 12.7mm thick	m	0.24	1.92	1.77	0.55	4.24
insulation board 19mm thick	m	0.22	1.76	2.34	0.61	4.71
insulation board 25mm thick	m	0.24	1.92	3.24	0.77	5.93

IRONMONGERY

	Unit	Labour hours	Net labour (£)	Net material (£)	O'heads /profit (£)	Total (£)
P21 IRONMONGERY						
Labour rate 8.00 per hour						
The material costs of ironmongery vary considerable due to wide variations in quality. The following rates are for fixing only.						
Fix only to softwood						
Casement stay and pin	nr	0.35	2.80	0.00	0.42	3.22
Casement fastener	nr	0.25	2.00	0.00	0.30	2.30
Hat and coat hook	nr	0.10	0.80	0.00	0.12	0.92
Shelf bracket	nr	0.35	2.80	0.00	0.42	3.22
Push plate	nr	0.15	1.55	0.00	0.23	1.78
Kicking plate	nr	0.20	1.60	0.00	0.24	1.84
Sliding door gear						
top track	m	0.30	2.40	0.00	0.36	2.76
bottom channel	m	0.30	2.40	0.00	0.36	2.76
close ends	nr	0.25	2.00	0.00	0.30	2.30
open bracket	nr	0.25	2.00	0.00	0.30	2.30
bottom guide	nr	0.25	2.00	0.00	0.30	2.30
door stop	nr	0.25	2.00	0.00	0.30	2.30
top runner	nr	0.33	2.64	0.00	0.40	3.04
Steel hinges						
light butts	nr	0.30	2.40	0.00	0.36	2.76
medium butts	nr	0.33	2.64	0.00	0.40	3.04
heavy butts	nr	0.35	2.80	0.00	0.42	3.22
rising butts	nr	0.40	3.20	0.00	0.48	3.68

Fix only ironmongery (cont'd)	Unit	Labour hours	Net labour (£)	Net material (£)	O'heads /profit (£)	Total (£)
Tee band hinges						
150 to 300mm	nr	0.80	6.40	0.00	0.96	7.36
350 to 600mm	nr	1.30	10.40	0.00	1.56	11.96
Barrel or tower bolts						
100 to 300mm	nr	0.55	4.40	0.00	0.66	5.06
350 to 450mm	nr	0.60	4.80	0.00	0.72	5.52
Helical door spring	nr	0.75	6.00	0.00	0.90	6.90
Overhead door spring						
medium	nr	1.00	8.00	0.00	1.20	9.20
heavy	nr	1.00	8.00	0.00	1.20	9.20
Door spring						
single action	nr	1.75	14.00	0.00	2.10	16.10
double action	nr	2.00	16.00	0.00	2.40	18.40
Sundries						
Postal knocker and letter plate	nr	1.00	8.00	0.00	1.20	9.20
Pull handles	nr	0.25	2.00	0.00	0.30	2.30
Flush pull handles	nr	0.40	3.20	0.00	0.48	3.68
Suffolk/Norfolk latch	nr	0.70	5.60	0.00	0.84	6.44
Hasp and staple	nr	0.25	2.00	0.00	0.30	2.30
Flush						
100 to 300mm	nr	1.20	9.60	0.00	1.44	11.04
300 to 450mm	nr	1.80	14.40	0.00	2.16	16.56
Indicating bolt	nr	0.60	4.80	0.00	0.72	5.52

IRONMONGERY

	Unit	Labour hours	Net labour (£)	Net material (£)	O'heads /profit (£)	Total (£)
Panic bolt						
single door	nr	2.20	17.60	0.00	2.64	20.24
double door	nr	3.40	27.20	0.00	4.08	31.28
Locks and latches						
cylinder rim night latch	nr	1.00	8.00	0.00	1.20	9.20
cylinder mortice night latch	nr	1.25	10.00	0.00	1.50	11.50
rim dead lock	nr	0.75	6.00	0.00	0.90	6.90
mortice dead lock	nr	1.00	8.00	0.00	1.20	9.20
rebated mortice lock	nr	1.20	9.60	0.00	1.44	11.04
mortice latch	nr	0.80	6.40	0.00	0.96	7.36
mortice sliding door lock	nr	0.90	7.20	0.00	1.08	8.28
mortice latch, furniture	nr	0.30	2.40	0.00	0.36	2.76
cupboard catch	nr	0.30	2.40	0.00	0.36	2.76

RATES FOR MEASURED WORK

	Unit	Labour hours	Net labour (£)	Net material (£)	O'heads /profit (£)	Total (£)
P31 HOLES/CHASES FOR SERVICES						
Labour rate 8.00 per hour						
Cutting chase in brickwork for pipe not exceeding 55mm nominal diameter	m	0.38	3.04	0.00	0.08	3.12
Cutting hole for pipe not exceeding 55mm nominal diameter						
Concrete						
100mm	nr	0.70	4.90	0.00	0.73	5.63
150mm	nr	1.10	7.70	0.00	1.16	8.86
200mm	nr	1.30	9.10	0.00	1.36	10.46
Reinforced concrete						
100mm	nr	1.60	11.20	0.00	1.68	12.88
150mm	nr	1.80	12.60	0.00	1.89	14.49
200mm	nr	2.00	14.00	0.00	2.10	16.10
Brickwork						
102mm	nr	0.40	2.80	0.00	0.42	3.22
215mm	nr	0.60	4.20	0.00	0.63	4.83
327mm	nr	0.90	6.30	0.00	0.94	7.24
Blockwork						
75mm	nr	0.35	2.45	0.00	0.37	2.82
100mm	nr	0.40	2.80	0.00	0.42	3.22
150mm	nr	0.50	3.50	0.00	0.53	4.03
215mm	nr	0.55	3.85	0.00	0.58	4.43
Softwood						
12mm	nr	0.04	0.32	0.00	0.05	0.37
25mm	nr	0.06	0.48	0.00	0.07	0.55
50mm	nr	0.08	0.64	0.00	0.10	0.74
100mm	nr	0.12	0.96	0.00	0.14	1.10
Hardwood						
12mm	nr	0.05	0.40	0.00	0.06	0.46
25mm	nr	0.07	0.56	0.00	0.08	0.64
50mm	nr	0.12	0.96	0.00	0.14	1.10
100mm	nr	0.18	1.44	0.00	0.22	1.66

Q

PAVING/FENCING

Building a better future

MATERIAL COSTS

	Unit	Material supply (£)	Material waste (%)	Total (£)
MATERIAL COSTS				
Granular fill	m3	12.00	1.0	12.12
Sand	m3	15.50	1.0	15.66
Hardcore	m3	8.20	1.0	8.28
Clinker	m3	12.00	1.0	12.12
Gravel				
20mm	t	11.50	10.0	12.65
40mm	t	11.30	10.0	12.43
Ready mix concrete mix A (1:3:6)	m3	48.59	5.0	51.02
Ready mix concrete mix B (1:2:4)	m3	56.44	5.0	59.26
Precast paving flags natural colour, size				
450 x 450 x 50mm	m2	10.89	5.0	11.43
600 x 600 x 50mm	m2	7.71	5.0	8.10
600 x 900 x 50mm	m2	6.56	5.0	6.89
Precast paving flags coloured, size				
450 x 450 x 50mm	m2	13.61	5.0	14.29
600 x 600 x 50mm	m2	9.64	5.0	10.12
600 x 900 x 50mm	m2	8.20	5.0	8.61
Charcon precast paving blocks thickness 65mm				
natural smooth	m2	7.07	5.0	7.42
red buff textured	m2	11.58	5.0	12.16
Precast concrete edgings				
51 x 152mm	m	1.51	5.0	1.59
51 x 203mm	m	1.79	5.0	1.88
51 x 254mm	m	2.15	5.0	2.26

RATES FOR MEASURED WORK

	Unit	Material supply (£)	Material waste (%)	Total (£)
Precast concrete kerbs				
Straight				
127 x 254mm	m	3.89	5.0	4.08
152 x 305mm	m	5.46	5.0	5.73
Curved				
127 x 254mm	m	5.51	5.0	5.79
152 x 305mm	m	7.77	5.0	8.16
Precast concrete channels				
straight, 127 x 254mm	m	3.63	5.0	3.81
curved, 127 x 254mm	m	4.90	5.0	5.15

CONCRETE EDGINGS/CHANNELS

	Unit	Labour hours	Net labour (£)	Net material (£)	O'heads /profit (£)	Total (£)
Q10 CONCRETE EDGINGS/CHANNELS						
Labour rate 8.00 per hour						
Excavate trench by hand to receive kerb foundation, size						
200 x 75mm	m	0.10	0.80	0.00	0.12	0.92
250 x 100mm	m	0.12	0.96	0.00	0.14	1.10
300 x 100mm	m	0.14	1.12	0.00	0.17	1.29
Excavate curved trench by hand to receive kerb foundation, size						
200 x 75mm	m	0.11	0.88	0.00	0.13	1.01
250 x 100mm	m	0.13	1.04	0.00	0.16	1.20
300 x 100mm	m	0.15	1.20	0.00	0.18	1.38
Ready mix concrete mix A in foundations for kerb, size						
200 x 75mm	m	0.01	0.08	0.76	0.13	0.97
250 x 100mm	m	0.02	0.16	1.28	0.22	1.66
300 x 100mm	m	0.03	0.24	1.53	0.27	2.04
Precast concrete kerbs, channels, edgings to BS5340, jointed and pointed in cement mortar						
Kerbs, straight						
127 x 254mm	m	0.40	3.20	4.03	1.08	8.31
152 x 305mm	m	0.40	3.20	5.73	1.34	10.27
Kerbs, curved						
127 x 254mm	m	0.50	4.00	5.51	1.43	10.94
152 x 305mm	m	0.50	4.00	7.77	1.77	13.54
Channels, straight						
127 x 254mm	m	0.40	3.20	3.63	1.02	7.85
Channels, curved						
127 x 254mm	m	0.50	4.00	4.90	1.33	10.24

RATES FOR MEASURED WORK

Precast concrete (cont'd)	Unit	Labour hours	Net labour (£)	Net material (£)	O'heads /profit (£)	Total (£)
Edgings						
51 x 152mm	m	0.30	2.40	1.59	0.60	4.59
51 x 203mm	m	0.30	2.40	1.88	0.64	4.92
51 x 254mm	m	0.30	2.40	2.26	0.70	5.36

HARDCORE SUB-BASES

	Unit	Labour hours	Net labour (£)	Net material (£)	O'heads /profit (£)	Total (£)
Q20 HARDCORE SUB-BASES						
Labour rate 8.00 per hour						
Beds and bases, compacting in layers where necessary, grading, thickness						
Average 100mm						
granular fill	m2	0.10	0.80	1.34	0.32	2.46
sand	m2	0.12	0.96	1.71	0.40	3.07
hardcore	m2	0.13	1.04	1.06	0.32	2.42
Average 150mm						
granular fill	m2	0.12	0.96	2.01	0.45	3.42
sand	m2	0.14	1.12	2.56	0.55	4.23
hardcore	m2	0.18	1.44	1.58	0.45	3.47
Average 200mm						
granular fill	m2	0.15	1.20	2.68	0.58	4.46
sand	m2	0.17	1.36	3.41	0.72	5.49
hardcore	m2	0.20	1.60	2.11	0.56	4.27

RATES FOR MEASURED WORK

	Unit	Labour hours	Net labour (£)	Net material (£)	O'heads /profit (£)	Total (£)
Q21 IN SITU CONCRETE BEDS						
Labour rate 8.00 per hour						
Site mixed concrete mix B in beds						
not exceeding 150mm thick	m3	2.80	22.40	59.26	12.25	93.91
150 to 450mm thick	m3	1.70	13.60	59.26	10.93	83.79
Formwork to sides of foundations						
not exceeding 250mm wide	m	0.75	6.00	0.96	1.04	8.00
200 to 500mm wide	m	1.28	10.24	1.91	1.82	13.97
Steel fabric reinforcement BS4483, laid in concrete beds						
A142	m2	0.12	0.96	1.62	0.39	2.97
A193	m2	0.15	1.20	2.21	0.51	3.92
Extra for placing concrete around reinforcement	m3	0.55	4.40	0.00	0.66	5.06
Expansion joint, Flexcell impregnated fibre based joint filler, formed joint 12.5mm thick						
not exceeding 150mm wide	m	0.12	0.96	1.94	0.43	3.34
150 to 300mm wide	m	0.18	1.44	2.82	0.64	4.90
Prepare level surfaces of unset concrete						
tamping by mechanical means	m2	0.06	0.48	0.00	0.07	0.55
power floating	m2	0.15	1.20	0.00	0.18	1.38
trowelling	m2	0.23	1.84	0.00	0.28	2.12

MACADAM PAVINGS

	Unit	Labour hours	Net labour (£)	Net material (£)	O'heads /profit (£)	Total (£)
Q22 MACADAM PAVINGS						
The following are specialist sub-contractor's prices and include 15% overheads and profit						
Bitumen macadam to BS4987, 50mm thick base course with 28mm aggregate, 20mm thick wearing course with 6mm aggregate, to pavings	m2	0.00	0.00	17.25	0.00	17.25
Bitumen macadam to BS4987, 70mm thick base course with 40mm aggregate, 25mm thick wearing course with 10mm aggregate, to pavings	m2	0.00	0.00	30.81	0.00	30.81

	Unit	Labour hours	Net labour (£)	Net material (£)	O'heads /profit (£)	Total (£)

Q23 GRAVEL PAVINGS

Labour rate 8.00 per hour

Gravel paving in two layers

	Unit	Labour hours	Net labour (£)	Net material (£)	O'heads /profit (£)	Total (£)
first layer clinker aggregate in bed 50mm thick	m2	0.08	0.64	0.63	0.19	1.46
first layer clinker aggregate in bed 75mm thick	m2	0.10	0.80	0.95	0.26	2.01
second layer 20mm gravel in bed 40mm thick	m2	0.03	0.24	0.51	0.11	0.86
second layer 20mm gravel in bed 60mm thick	m2	0.03	0.24	0.76	0.15	1.15
second layer 40mm gravel in bed 50mm thick	m2	0.04	0.32	0.62	0.14	1.08
second layer 40mm gravel in bed 70mm thick	m2	0.04	0.32	0.87	0.18	1.37

BRICK PAVINGS

	Unit	Labour hours	Net labour (£)	Net material (£)	O'heads /profit (£)	Total (£)

Q24 BRICK PAVINGS

Labour rate 8.00 per hour

Brick paviours, 215 x 103 x 65mm
(PC 400/1000), laid to falls and
cross jointed with 15mm thick
cement mortar (1:3)

The areas over 300mm wide,
straight joints both ways

bricks laid flat	m2	1.00	8.00	18.50	3.97	30.48
bricks laid on edge	m2	1.25	10.00	29.30	5.89	45.19

To areas over 300mm wide, herring-
bone pattern

bricks laid flat	m2	1.25	10.00	18.50	4.27	32.78
bricks laid on edge	m2	1.50	12.00	29.30	6.19	47.50

Add/deduct for variation of 5
pounds in price per 1000 of brick

bricks laid flat	m	0.00	0.00	2.25	0.34	2.59
bricks laid on edge	m	0.00	0.00	3.55	0.53	4.08

	Unit	Labour hours	Net labour (£)	Net material (£)	O'heads /profit (£)	Total (£)
Q25 SETT PAVINGS						
Labour rate 8.00 per hour						
Precast concrete paving flags (BS368), spot bedded in cement lime mortar (1:1:6), size						
Natural colour						
450 x 450 x 50mm	m2	0.43	3.44	11.43	2.23	17.10
600 x 600 x 50mm	m2	0.34	2.72	8.10	1.62	12.44
600 x 900 x 50mm	m2	0.31	2.48	6.89	1.41	10.78
Coloured, size						
450 x 450 x 50mm	m2	0.43	3.44	14.29	2.66	20.39
600 x 600 x 50mm	m2	0.34	2.72	10.12	1.93	14.77
600 x 900 x 50mm	m2	0.31	2.48	8.61	1.66	12.75
Charcon precast concrete blocks size 200 x 100 x 65mm thick, laid flat on sand bed 50mm thick						
natural smooth	m2	0.45	3.60	7.42	1.65	12.67
red buff textured	m2	0.45	3.60	12.16	2.36	18.12

FENCING

	Unit	Labour hours	Net labour (£)	Net material (£)	O'heads /profit (£)	Total (£)

Q40 FENCING

The following are specialist sub-contractor's prices and include 15% overheads and profit

Chainlink fencing

Chainlink fencing (BS1722 Part 1), galvanized steel mesh, three strained line wires, concrete posts at 3m centres, height

1.20m	m	0.00	0.00	15.00	0.00	15.00
1.80m	m	0.00	0.00	18.00	0.00	18.00

Chainlink fencing (BS1722 Part 1), plastic coated mesh, three strained line wires, concrete posts at 3m centres

1.20m	m	0.00	0.00	16.50	0.00	16.50
1.80m	m	0.00	0.00	19.00	0.00	19.00

Extra for one line of barbed wire
to top line wire	m	0.00	0.00	0.15	0.00	0.15

Strained wire fence (BS1722 Part 3), concrete intermediate posts at 3m centres, holed for wire, height

1.00m (5 wires)	m	0.00	0.00	10.00	0.00	10.00
1.20m (6 wires)	m	0.00	0.00	10.30	0.00	10.30
1.40m (8 wires)	m	0.00	0.00	11.00	0.00	11.00

Fencing (cont'd)	Unit	Labour hours	Net labour (£)	Net material (£)	O'heads /profit (£)	Total (£)
Cleft chestnut fencing (BS1722 Part 4), timber posts average 2.5m centres, pales average 75mm apart, height						
0.90m	m	0.00	0.00	8.50	0.00	8.50
1.05m	m	0.00	0.00	8.00	0.00	8.00
1.20m	m	0.00	0.00	9.50	0.00	9.50
1.35m	m	0.00	0.00	15.00	0.00	15.00
1.80m	m	0.00	0.00	18.00	0.00	18.00
Close boarded fence (BS1722 Part 5), pales lapped at 13mm centres, treated softwood intermediate posts at 3m centres, height						
1.00m	m	0.00	0.00	26.00	0.00	26.00
1.20m	m	0.00	0.00	27.00	0.00	27.00
1.40m	m	0.00	0.00	36.00	0.00	36.00
1.60m	m	0.00	0.00	37.00	0.00	37.00
1.80m	m	0.00	0.00	40.00	0.00	40.00
Wooden palisade fence (BS1722 Part 6), pales spaced 75mm apart, intermediate concrete posts at 3m centres, height						
1.00m	m	0.00	0.00	20.00	0.00	20.00
1.20m	m	0.00	0.00	21.00	0.00	21.00
1.40m	m	0.00	0.00	22.00	0.00	22.00
1.60m	m	0.00	0.00	25.00	0.00	25.00
1.80m	m	0.00	0.00	26.50	0.00	26.50

R

DISPOSAL SYSTEMS

R10 Rainwater goods

R11 Foul drains above ground

R12 Drainage below ground

MATERIAL COSTS

	Unit	Material supply (£)	Material waste (%)	Total (£)

MATERIAL COSTS

The following allowances have been made for waste in this section:

Rainwater pipes - 5%

Rainwater gutters - 5%

Rainwater fittings - 2.5%

Waster pipes - 7.5%

Waste fittings - 5%

Soil pipes - 7.5%

Soil fittings - 5%

Drain fittings - 5%

	Unit	Labour hours	Net labour (£)	Net material (£)	O'heads /profit (£)	Total (£)

R10 RAINWATER GOODS

PIPES

Labour rate 9.60 per hour

uPVC Terrain 110 system rainwater pipes, straight half-round, solvent-welded coupling joints to brickwork with uPVC holderbats at 2m maximum centres, plugging

	Unit	Labour hours	Net labour (£)	Net material (£)	O'heads /profit (£)	Total (£)
100mm diameter pipe	m	0.37	3.55	11.56	2.27	17.38
Extra over for						
shoe	nr	0.39	3.74	8.57	1.85	14.16
bend, 92.5 degrees	nr	0.34	3.26	9.67	1.94	14.87
bend, 112.5 degrees	nr	0.34	3.26	9.67	1.94	14.87
bend, 135 degrees	nr	0.34	3.26	9.67	1.94	14.87
access bend, 92.5 degrees	nr	0.34	3.26	21.68	3.74	28.68
branch, 92.5 degrees	nr	0.39	3.74	12.64	2.46	18.84
branch, 135 degrees	nr	0.39	3.74	12.78	2.48	19.00
access branch, 92.5 degrees	nr	0.39	3.74	24.76	4.28	32.78
spigot socket bend	nr	0.34	3.26	7.98	1.69	12.93
seal ring adaptor	nr	0.23	2.21	1.45	0.55	4.21
access pipe, solvent welded socket	nr	0.34	3.26	12.06	2.30	17.62
access pipe, spigot and socket	nr	0.34	3.26	11.85	2.27	17.38
connection to drain pipe, caulking bush	nr	0.30	2.88	5.99	1.33	10.20

uPVC Marley circular system, straight, loose spigot and socket joints, to brickwork with pipe clips at 2m maximum centres, plugging

	Unit	Labour hours	Net labour (£)	Net material (£)	O'heads /profit (£)	Total (£)
51mm diameter Miniline pipe	m	0.22	2.11	3.97	0.91	6.99
Extra over for						
offset, 76mm	nr	0.15	1.44	3.77	0.78	5.99
offset, 152mm	nr	0.15	1.44	4.08	0.83	6.35
offset bends, 67.5 degrees	nr	0.15	1.44	1.47	0.44	3.35
shoe	nr	0.30	2.88	1.47	0.65	5.00

RAINWATER GOODS

	Unit	Labour hours	Net labour (£)	Net material (£)	O'heads /profit (£)	Total (£)
68mm diameter	m	0.25	2.40	4.20	0.99	7.59
Extra over for						
bend, 45 degrees	nr	0.15	1.44	3.09	0.68	5.21
bend, 67.5 degrees	nr	0.15	1.44	3.09	0.68	5.21
bend, 87.5 degrees	nr	0.15	1.44	3.09	0.68	5.21
offset bend, 67.5 degrees	nr	0.15	1.44	1.71	0.47	3.62
offset bend, 20 degrees	nr	0.15	1.44	1.86	0.49	3.79
branch, 67.5 degrees	nr	0.15	1.44	6.16	1.14	8.74
access pipe	nr	0.25	2.40	8.46	1.63	12.49
reducer, 110 to 68mm	nr	0.15	1.44	2.65	0.61	4.70
reducer, 82 to 68mm	nr	0.15	1.44	2.17	0.54	4.15
shoe	nr	0.30	2.88	2.65	0.83	6.36
65mm square pipe	m	0.25	2.40	4.98	0.18	7.56
Extra over for						
bend, 67.5 degrees	nr	0.15	1.44	2.00	0.52	3.96
bend, 87.5 degrees	nr	0.15	1.44	2.29	0.56	4.29
offset, 54mm	nr	0.15	1.44	3.49	0.12	5.05
branch, 67.5 degrees	nr	0.15	1.44	7.68	1.37	10.49
access pipe	nr	0.25	2.40	10.64	1.96	15.00
shoe	nr	0.30	2.88	2.28	0.77	5.93
drain adaptor	nr	0.15	1.44	1.70	0.47	3.61
82mm diameter pipe	m	0.28	2.69	9.42	1.82	13.93
Extra over for						
bend, 87.5 degrees	nr	0.18	1.73	6.74	1.27	9.74
bend, adjustable	nr	0.18	1.73	11.32	1.96	15.01
offset bend, socket 67.5 degrees	nr	0.18	1.73	6.74	1.27	9.74
offset bend, spigot 67.5 degrees	nr	0.18	1.73	5.71	1.12	8.56
branch, adjustable	nr	0.18	1.73	10.58	1.85	14.16
access pipe	nr	0.28	2.69	12.98	2.35	18.02
shoe	nr	0.32	3.07	7.57	1.60	12.24
adaptor	nr	0.15	1.44	4.37	0.87	6.68

319

RATES FOR MEASURED WORK

Rainwater pipes (cont'd)	Unit	Labour hours	Net labour (£)	Net material (£)	O'heads /profit (£)	Total (£)
110mm diameter pipe	m	0.30	2.88	9.58	1.87	14.33
Extra over for						
bend, 87.5 degrees	nr	0.20	1.92	8.92	1.63	12.47
bend, adjustable	nr	0.20	1.92	11.32	1.99	15.23
offset bend, end socket						
67.5 degrees	nr	0.20	1.92	7.50	1.41	10.83
branch, 87.5 degrees	nr	0.20	1.92	12.54	2.17	16.63
access pipe	nr	0.25	2.40	14.19	2.49	19.08
shoe	nr	0.34	3.26	9.79	1.96	15.01
adaptor	nr	0.15	1.44	3.67	0.77	5.88
160mm diameter pipe	m	0.35	3.36	21.07	3.66	28.09
Extra over for						
bend, adjustable	nr	0.24	2.30	22.44	3.71	28.45
offset bend, spigot 67.5						
degrees	nr	0.24	2.30	22.44	3.71	28.45
offset bend, spigot 67.5						
degrees	nr	0.24	2.30	19.61	0.55	22.46
branch, 87.5 degrees	nr	0.24	2.30	35.11	5.61	43.02
access pipe	nr	0.28	2.69	28.59	4.69	35.97
shoe	nr	0.40	3.84	20.87	3.71	28.42
adaptor	nr	0.15	1.44	11.70	1.97	15.11

uPVC, Osma Minifit system, straight, connector fixing bracket joints, to brickwork with brackets at 2m maximum centres, plugging

55mm square pipe	m	0.20	1.92	2.34	0.64	4.90
Extra over for						
bend, 45 degrees	nr	0.14	1.34	1.73	0.46	3.53
bend, 87.5 degrees	nr	0.14	1.34	2.01	0.50	3.85
offset bend	nr	0.14	1.34	1.29	0.39	3.02
shoe	nr	0.18	1.73	1.29	0.45	3.47

320

RAINWATER GOODS

	Unit	Labour hours	Net labour (£)	Net material (£)	O'heads /profit (£)	Total (£)
uPVC, Osma Roundline system, straight, connector joints, to brickwork with pipe brackets at 2m centres, plugging						
68mm diameter pipe	m	0.25	2.40	2.84	0.79	6.03
Extra over for						
bend, 87.5 degrees	nr	0.15	1.44	2.79	0.63	4.86
offset bend	nr	0.15	1.44	1.56	0.45	3.45
shoe	nr	0.30	2.88	1.64	0.68	5.20
branch, 67.5 degrees	nr	0.20	1.92	4.94	1.03	7.89
access pipe	nr	0.15	1.44	6.79	1.23	9.46
universal connector	nr	0.20	1.92	2.57	0.67	5.16
uPVC, Hepworth system, straight, connector joints, to brickwork at 2m maximum centres, plugging						
50mm diameter pipe	m	0.22	2.11	3.47	0.84	6.42
Extra over for						
offset bend, 112.5 degrees	nr	0.15	1.44	2.02	0.52	3.98
shoe	nr	0.30	2.88	1.98	0.73	5.59
68mm diameter pipe	m	0.25	2.40	3.42	0.87	6.69
Extra over for						
offset bend, 92.5 degrees	nr	0.15	1.44	2.96	0.66	5.06
offset bend, 112.5 degrees	nr	0.15	1.44	1.51	0.44	3.39
branch, 112.5 degrees	nr	0.15	1.44	5.38	1.02	7.84
shoe	nr	0.30	2.88	1.51	0.66	5.05
access pipe	nr	0.15	1.44	7.06	1.27	9.77
65mm square pipe	m	0.20	1.92	4.12	0.15	6.19
Extra over for						
shoe	nr	0.18	1.73	2.29	0.60	4.62
offset bend, 112.5 degrees	nr	0.14	1.34	1.70	0.46	3.50
mini offset, 44mm	nr	0.14	1.34	4.13	0.82	6.29
adaptor, square to round	nr	0.14	1.34	2.53	0.58	4.45
branch, 112.5 degrees	nr	0.14	1.34	5.47	1.02	7.83
access pipe	nr	0.14	1.34	6.73	1.21	9.28

RATES FOR MEASURED WORK

	Unit	Labour hours	Net labour (£)	Net material (£)	O'heads /profit (£)	Total (£)
GUTTERS						
uPVC, Terrain 2100 system, straight half-round, joint bracket joints, to timber with support brackets at 1m maximum centres						
110mm gutter	m	0.26	2.50	3.60	0.91	7.02
Extra over for						
running outlet	nr	0.26	2.50	3.01	0.83	6.34
angle, 90 degrees	nr	0.26	2.50	3.04	0.83	6.37
angle, 120 degrees	nr	0.26	2.50	3.61	0.92	7.03
angle, 135 degrees	nr	0.26	2.50	3.14	0.85	6.49
short stop end	nr	0.14	1.34	1.49	0.42	3.25
long stop end	nr	0.14	1.34	2.91	0.64	4.89
uPVC adaptor to cast-iron half-round	nr	0.28	2.69	4.40	1.06	8.15
aluminium adaptor to cast-iron half-round	nr	0.28	2.69	16.69	2.91	22.29
uPVC adaptor to cast-iron ogee	nr	0.28	2.69	4.46	1.07	8.22
uPVC, Marley Miniline system, gutter union joints, to timber with support brackets at 1m maximum centres						
75mm half-round gutter	m	0.26	2.50	1.82	0.65	4.97
Extra over for						
angle, 45 degrees	nr	0.26	2.50	2.37	0.73	5.60
angle, 90 degrees	nr	0.26	2.50	2.37	0.73	5.60
outlet	nr	0.26	2.50	1.99	0.67	5.16
stop end, internal	nr	0.14	1.34	0.86	0.33	2.53
stop end, external	nr	0.16	1.54	0.86	0.36	2.76

RAINWATER GOODS

	Unit	Labour hours	Net labour (£)	Net material (£)	O'heads /profit (£)	Total (£)
uPVC, Marley Deepflow system, gutter union joints, to timber with support brackets at 1m maximum centres						
110mm semi-elliptical	m	0.30	2.88	3.98	1.03	7.89
Extra over for						
angle, 45 degrees	nr	0.28	2.69	3.63	0.95	7.27
angle, 90 degrees	nr	0.28	2.69	5.44	1.22	9.35
outlet	nr	0.18	1.73	3.32	0.76	5.81
stop end	nr	0.18	1.73	1.77	0.53	4.03
stop end, outlet	nr	0.18	1.73	2.91	0.12	4.76
uPVC, Marley Premier system, gutter union joints, to timber with support brackets at 1m maximum centres						
112mm half-round gutter	m	0.26	2.50	3.57	0.91	6.98
Extra over for						
angle, 45 degrees	nr	0.26	2.50	3.91	0.96	7.37
angle, 90 degrees	nr	0.26	2.50	3.72	0.93	7.15
outlet	nr	0.26	2.50	3.40	0.89	6.79
stop end	nr	0.14	1.34	3.40	0.71	5.45
stop end, outlet	nr	0.16	1.54	3.40	0.74	5.68
adaptor to ogee	nr	0.28	2.69	9.65	1.85	14.19
uPVC, Osma Minifit system, joint bracket joints, to timber with support brackets at 1m maximum centres						
75mm half-round gutter	m	0.24	2.30	1.91	0.63	4.84
Extra over for						
running outlet	nr	0.24	2.30	1.69	0.60	4.59
angle, 45 degrees	nr	0.24	2.30	1.86	0.62	4.78
angle, 90 degrees	nr	0.24	2.30	1.65	0.59	4.54
stop end	nr	0.14	1.34	0.74	0.31	2.39
stop end, outlet	nr	0.14	1.34	1.65	0.07	3.06

RATES FOR MEASURED WORK

Gutters (cont'd)	Unit	Labour hours	Net labour (£)	Net material (£)	O'heads /profit (£)	Total (£)
uPVC, Osma Roundline system, joint bracket joints, to timber with support brackets at 1m maximum centres						
112mm half-round gutter	m	0.26	2.50	2.95	0.82	6.27
Extra over for						
swivelock running outlet	nr	0.26	2.50	2.72	0.78	6.00
swivelock running outlet with offset bend	nr	0.26	2.50	3.79	0.94	7.23
angle, 45 degrees	nr	0.26	2.50	2.82	0.80	6.12
angle, 60 degrees	nr	0.26	2.50	2.82	0.80	6.12
angle, 90 degrees	nr	0.26	2.50	2.55	0.76	5.81
stop end, internal	nr	0.14	1.34	0.69	0.30	2.33
stop end, external	nr	0.14	1.34	1.10	0.37	2.81
connector to half-round cast-iron gutter	nr	0.28	2.69	2.24	0.74	5.67
connector to ogee cast-iron gutter	nr	0.28	2.69	3.56	0.94	7.19
uPVC, Hepworth system, straight, joint bracket joints, to timber with support brackets at 1m maximum centres						
76mm half-round gutter	m	0.26	2.50	2.16	0.70	5.36
Extra over for						
stop end, outlet	nr	0.14	1.34	2.33	0.09	3.76
running outlet	nr	0.26	2.50	2.33	0.72	5.55
stop end	nr	0.14	1.34	0.99	0.35	2.68
angle, 90 degrees	nr	0.26	2.50	3.37	0.88	6.75
112mm half-round gutter	m	0.26	2.50	2.82	0.80	6.12
Extra over for						
running outlet	nr	0.26	2.50	2.04	0.68	5.22
stop end	nr	0.14	1.34	1.14	0.37	2.85
angle, 90 degrees	nr	0.26	2.50	2.35	0.73	5.58
angle, 120 degrees	nr	0.26	2.50	2.69	0.78	5.97
angle, 135 degrees	nr	0.26	2.50	2.69	0.78	5.97
adaptor to metal ogee	nr	0.26	2.50	4.76	1.09	8.35

RAINWATER GOODS

	Unit	Labour hours	Net labour (£)	Net material (£)	O'heads /profit (£)	Total (£)
150mm half-round gutter	m	0.30	2.88	6.88	1.46	11.22
Extra over for						
stop end, outlet	nr	0.16	1.54	7.08	0.22	8.84
running outlet	nr	0.30	2.88	7.37	1.54	11.79
stop end	nr	0.16	1.54	3.37	0.74	5.65
angle, 90 degrees	nr	0.30	2.88	9.72	1.89	14.49
112mm square system gutter	m	0.26	2.50	2.91	0.81	6.22
Extra over for						
running outlet	nr	0.26	2.50	2.97	0.82	6.29
stop end	nr	0.14	1.34	1.05	0.36	2.75
angle, 90 degrees	nr	0.26	2.50	2.88	0.81	6.19
angle, 120 degrees	nr	0.26	2.50	3.65	0.92	7.07
angle, 135 degrees	nr	0.26	2.50	3.65	0.92	7.07
115 x 75mm high-capacity gutter	m	0.30	2.88	3.29	0.93	7.10
Extra over for						
running outlet	nr	0.30	2.88	3.44	0.95	7.27
stop end	nr	0.16	1.54	1.64	0.48	3.66
angle, 90 degrees	nr	0.30	2.88	3.34	0.93	7.15
angle, 135 degrees	nr	0.30	2.88	4.97	1.18	9.03
adaptor to deep section gutter	nr	0.30	2.88	2.69	0.14	5.71
adaptor to square down pipe	nr	0.30	2.88	2.44	0.80	6.12

RATES FOR MEASURED WORK

	Unit	Labour hours	Net labour (£)	Net material (£)	O'heads /profit (£)	Total (£)
ANCILLARIES						
Roof and balcony outlets						
uPVC rainwater heads						
Terrain						
249 x 131 x 203mm	nr	0.40	3.84	8.18	1.80	13.82
280 x 121 x 194mm	nr	0.50	4.80	8.10	1.93	14.83
254 x 180 x 221mm	nr	0.50	4.80	25.31	4.52	34.63
254 x 180 x 222mm	nr	0.50	4.80	16.02	3.12	23.94
Marley						
280 x 155 x 230mm	nr	0.50	4.80	9.89	2.20	16.89
280 x 230 x 155mm	nr	0.50	4.80	11.85	2.50	19.15
406 x 375 x 248mm	nr	0.70	6.72	51.16	8.68	66.56
406 x 375 x 248mm	nr	0.70	6.72	51.16	8.68	66.56
Osma						
254 x 178 x 145mm	nr	0.40	3.84	6.31	1.52	11.67
Hepworth						
267 x 140 x 114mm	nr	0.50	4.80	8.21	1.95	14.96

	Unit	Labour hours	Net labour (£)	Net material (£)	O'heads /profit (£)	Total (£)
R11 FOUL DRAINS ABOVE GROUND						
WASTE PIPES						
MuPVC Terrain waste system 200, solvent-welded joints, to brickwork with clips at 500mm maximum centres, plugging						
32mm diameter pipe	m	0.24	2.30	2.44	0.71	5.45
Extra over for						
bend, 91.25 degrees	nr	0.22	2.11	1.86	0.60	4.57
bend, 135 degrees	nr	0.22	2.11	1.86	0.60	4.57
bend, 165 degrees	nr	0.22	2.11	1.97	0.61	4.69
tee, 91.25 degrees	nr	0.25	2.40	2.65	0.76	5.81
tee, 135 degrees	nr	0.25	2.40	2.85	0.79	6.04
connector, iron to MuPVC	nr	0.22	2.11	1.59	0.55	4.25
connector, copper to MuPVC	nr	0.22	2.11	2.10	0.63	4.84
connection to back inlet gulley, caulking bush	nr	0.15	1.44	2.22	0.55	4.21
40mm diameter pipe	m	0.27	2.59	2.99	0.84	6.42
Extra over for						
bend, 91.25 degrees	nr	0.24	2.30	2.10	0.66	5.06
bend, 135 degrees	nr	0.24	2.30	2.10	0.66	5.06
bend, 165 degrees	nr	0.24	2.30	2.32	0.69	5.31
tee, 91.25 degrees	nr	0.28	2.69	3.30	0.90	6.89
tee, 135 degrees	nr	0.28	2.69	3.44	0.92	7.05
sweep cross	nr	0.32	3.07	9.48	1.88	14.43
connector, iron to MuPVC	nr	0.24	2.30	1.87	0.63	4.80
connector, copper to MuPVC	nr	0.24	2.30	2.22	0.68	5.20
connection to back inlet gulley, caulking bush	nr	0.17	1.63	1.59	0.48	3.70
50mm diameter pipe	m	0.30	2.88	4.39	1.09	8.36
Extra over for						
bend, 91.25 degrees	nr	0.26	2.50	3.02	0.83	6.35
bend, 135 degrees	nr	0.26	2.50	3.11	0.84	6.45
bend, 165 degrees	nr	0.26	2.50	3.29	0.87	6.66
tee, 91.25 degrees	nr	0.30	2.88	5.75	1.29	9.92
tee, 135 degrees	nr	0.30	2.88	6.01	1.33	10.22

	Unit	Labour hours	Net labour (£)	Net material (£)	O'heads /profit (£)	Total (£)
sweep cross	nr	0.32	3.07	9.85	1.94	14.86
connector, iron to MuPVC	nr	0.26	2.50	2.68	0.78	5.96
connector, copper to MuPVC	nr	0.26	2.50	2.97	0.82	6.29
connection to back inlet						
gulley, caulking bush	nr	0.20	1.92	2.22	0.62	4.76

MuPVC, Marley waste system, push-fit joints, to brickwork with clips at 500mm maximum centres, plugging

32mm diameter pipe	m	0.21	2.02	0.67	0.40	3.09
Extra over for						
bend, 90 degrees	nr	0.19	1.82	0.78	0.39	2.99
bend, 45 degrees	nr	0.19	1.82	1.86	0.55	4.23
branch, 88.5 degrees	nr	0.22	2.11	2.09	0.63	4.83
tank connector	nr	0.20	1.92	1.19	0.47	3.58
multi-fit waste connector	nr	0.20	1.92	1.52	0.52	3.96
40mm diameter pipe	m	0.21	2.02	0.84	0.43	3.29
Extra over for						
bend, 90 degrees	nr	0.19	1.82	0.81	0.39	3.02
bend, 45 degrees	nr	0.19	1.82	0.87	0.40	3.09
branch, 88.5 degrees	nr	0.22	2.11	1.30	0.51	3.92
reducer, 40 to 32mm	nr	0.22	2.11	0.87	0.45	3.43
tank connector	nr	0.20	1.92	1.26	0.48	3.66
multi-fit waste connector	nr	0.20	1.92	1.58	0.53	4.03

Polypropylene, Hepworth waste systems, ring-seal connector joints, to brickwork with 500mm maximum centres, plugging

32mm diameter pipe	m	0.21	2.02	0.68	0.41	3.11
Extra over for						
reducer, 32 to 21.5mm	nr	0.19	1.82	0.45	0.34	2.61
tank connector, to iron	nr	0.19	1.82	0.60	0.36	2.78
tank connector, to copper	nr	0.19	1.82	0.60	0.06	2.48
bend, 135 degrees	nr	0.19	1.82	0.60	0.36	2.78
bend, 150 degrees	nr	0.19	1.82	0.60	0.36	2.78

FOUL DRAINS ABOVE GROUND

Hepworth waste pipes (cont'd)	Unit	Labour hours	Net labour (£)	Net material (£)	O'heads /profit (£)	Total (£)
bend, 90 degrees	nr	0.19	1.82	0.60	0.36	2.78
bend, 92.5 degrees	nr	0.19	1.82	0.60	0.36	2.78
tee, 92.5 degrees	nr	0.22	2.11	0.60	0.41	3.12
plug, access	nr	0.10	0.96	0.60	0.23	1.79
40mm diameter pipe	m	0.24	2.30	0.81	0.47	3.58
Extra over for						
reducer, 40 to 21.5mm	nr	0.21	2.02	0.45	0.37	2.84
reducer, 40 to 32mm	nr	0.21	2.02	0.60	0.39	3.01
tank connector, to iron	nr	0.21	2.02	0.60	0.39	3.01
tank connector, to copper	nr	0.21	2.02	0.60	0.39	3.01
bend, 135 degrees	nr	0.21	2.02	0.60	0.39	3.01
bend, 150 degrees	nr	0.21	2.02	0.60	0.39	3.01
bend, 90 degrees	nr	0.21	2.02	0.60	0.39	3.01
bend, 92.5 degrees	nr	0.21	2.02	0.60	0.39	3.01
tee, 92.5 degrees	nr	0.25	2.40	0.60	0.45	3.45
plug, access	nr	0.10	0.96	0.60	0.23	1.79
50mm diameter pipe	m	0.27	2.59	1.35	0.59	4.53
Extra over for						
reducer, 50 to 40mm	nr	0.23	2.21	1.07	0.49	3.77
reducer, 50 to 32mm	nr	0.23	2.21	1.07	0.49	3.77
bend, 135 degrees	nr	0.23	2.21	1.07	0.49	3.77
bend, 150 degrees	nr	0.23	2.21	1.07	0.49	3.77
bend, 90 degrees	nr	0.23	2.21	1.07	0.49	3.77
bend, 92.5 degrees	nr	0.23	2.21	1.07	0.49	3.77
tee, 92.5 degrees	nr	0.27	2.59	1.07	0.55	4.21

OVERFLOWS

uPVC, Terrain overflow system, solvent-welded joints, to brickwork with clips at 500mm maximum centres, plugging

	Unit	Labour hours	Net labour (£)	Net material (£)	O'heads /profit (£)	Total (£)
19mm diameter pipe	m	0.19	1.82	1.01	0.42	3.25
Extra over for						
bend, 91 degrees	nr	0.17	1.63	0.90	0.38	2.91
bend, 135 degrees	nr	0.17	1.63	1.05	0.40	3.08
branch, 91 degrees	nr	0.20	1.92	1.05	0.45	3.42

329

	Unit	Labour hours	Net labour (£)	Net material (£)	O'heads /profit (£)	Total (£)
connector, bent tank	nr	0.20	1.92	1.44	0.50	3.86
connector, straight tank	nr	0.20	1.92	1.28	0.48	3.68
connector, iron to uPVC	nr	0.20	1.92	1.19	0.47	3.58

uPVC, Marley overflow system, solvent-welded connector joints, to brickwork with clips at 500mm maximum centres, plugging

22mm diameter pipe	m	0.19	1.82	0.90	0.41	3.13
Extra over for						
bend, 88.5 degrees	nr	0.17	1.63	0.87	0.37	2.87
bend, adjustable	nr	0.17	1.63	1.18	0.42	3.23
branch, 88.5 degrees	nr	0.20	1.92	0.80	0.41	3.13
adaptor, iron to uPVC	nr	0.17	1.63	0.74	0.36	2.73
straight tank connector	nr	0.20	1.92	1.25	0.48	3.65
bent tank connector	nr	0.20	1.92	1.32	0.49	3.73

Polypropylene, Osma/ClearBore, ring-sealed socket joints, to brickwork with clips at 500mm maximum centres, plugging

19mm diameter pipe	m	0.19	1.82	0.45	0.34	2.61
Extra over for						
bend, 90 degrees	nr	0.17	1.63	0.57	0.33	2.53
tee, 90 degrees	nr	0.20	1.92	0.67	0.39	2.98
connector, bent tank	nr	0.20	1.92	0.87	0.42	3.21
connector, straight tank	nr	0.20	1.92	0.87	0.42	3.21

FOUL DRAINS ABOVE GROUND

	Unit	Labour hours	Net labour (£)	Net material (£)	O'heads /profit (£)	Total (£)
Polypropylene, Hepworth Push-Fit overflow system, ring-sealed socket with clips at 500mm maximum centres, plugging						
19mm diameter pipe	m	0.19	1.82	0.37	0.33	2.52
Extra over for						
bend, 90 degrees	nr	0.17	1.63	0.45	0.31	2.39
tee, 90 degrees	nr	0.20	1.92	0.45	0.36	2.73
reducer	nr	0.17	1.63	0.45	0.31	2.39
connector, straight tank	nr	0.20	1.92	0.45	0.36	2.73
connector, bent tank	nr	0.20	1.92	0.45	0.36	2.73

TRAPS

Polypropylene, Terrain, screwed joint to outlet and pipe

Bottle P trap

32mm	nr	0.27	2.59	4.27	1.03	7.89
40mm	nr	0.32	3.07	5.11	1.23	9.41

Bottle S trap

32mm	nr	0.27	2.59	5.16	1.16	8.91
40mm	nr	0.32	3.07	6.27	1.40	10.74

Self-resealing bottle P trap

32mm	nr	0.27	2.59	5.32	1.19	9.10
40mm	nr	0.32	3.07	6.24	1.40	10.71

Self-resealing bottle S trap

32mm	nr	0.27	2.59	6.29	1.33	10.21
40mm	nr	0.32	3.07	7.29	1.55	11.91

Telescopic bottle P trap

32mm	nr	0.27	2.59	7.28	1.48	11.35
40mm	nr	0.32	3.07	8.75	1.77	13.59

Traps (cont'd)	Unit	Labour hours	Net labour (£)	Net material (£)	O'heads /profit (£)	Total (£)
Tubular P trap						
32mm	nr	0.27	2.59	3.84	0.96	7.39
40mm	nr	0.32	3.07	4.43	1.12	8.62
50mm	nr	0.35	3.36	10.12	2.02	15.50
Tubular S trap						
32mm	nr	0.27	2.59	4.87	1.12	8.58
40mm	nr	0.32	3.07	5.70	1.32	10.09
Washing machine trap						
40mm	nr	0.40	3.84	7.71	1.73	13.28
Tubular running P trap						
32mm	nr	0.30	2.88	5.90	1.32	10.10
40mm	nr	0.32	3.07	6.44	1.43	10.94
50mm	nr	0.35	3.36	14.84	2.73	20.93
Tubular P trap for bath						
40mm	nr	0.33	3.17	8.24	1.71	13.12
Tubular P trap assembly for bath						
40mm, white	nr	0.50	4.80	14.05	2.83	21.68
40mm, chromium-plated	nr	0.50	4.80	16.84	3.25	24.89
Tubular P lowline bath trap						
40mm	nr	0.50	4.80	4.82	1.44	11.06
Tubular P lowline bath trap with overflow						
40mm	nr	0.50	4.80	7.43	1.83	14.06

FOUL DRAINS ABOVE GROUND

	Unit	Labour hours	Net labour (£)	Net material (£)	O'heads /profit (£)	Total (£)
Polypropylene, Marley, screwed joint to outlet and pipe						
Tubular P trap						
32mm	nr	0.27	2.59	2.76	0.80	6.15
40mm	nr	0.32	3.07	3.32	0.96	7.35
Tubular S trap						
32mm	nr	0.27	2.59	3.32	0.89	6.80
40mm	nr	0.32	3.07	3.95	1.05	8.07
Standard bottle trap						
32mm	nr	0.25	2.40	2.84	0.79	6.03
40mm	nr	0.30	2.88	2.66	0.83	6.37
Monitor anti-syphon bottle trap						
32mm	nr	0.25	2.40	4.49	1.03	7.92
40mm	nr	0.30	2.88	4.69	1.14	8.71
Low inlet tubular P trap, 40mm						
plain	nr	0.33	3.17	3.32	0.97	7.46
with access	nr	0.33	3.17	3.87	1.06	8.10
with overflow and outlet	nr	0.50	4.80	9.57	2.16	16.53

333

RATES FOR MEASURED WORK

Marley traps (cont'd)	Unit	Labour hours	Net labour (£)	Net material (£)	O'heads /profit (£)	Total (£)
shallow bath trap, 40mm	nr	0.33	3.17	4.18	1.10	8.45
low level bath P trap	nr	0.35	3.36	3.91	1.09	8.36
washing machine kit	nr	0.40	3.84	5.60	1.42	10.86
Running P trap						
40mm	nr	0.33	3.17	4.76	1.19	9.12
50mm	nr	0.33	3.17	11.25	2.16	16.58
Polypropylene, Hepworth, screwed joint to outlet and pipe						
Bottle trap with 38mm seal						
32mm	nr	0.25	2.40	1.99	0.66	5.05
40mm	nr	0.30	2.88	2.30	0.78	5.96
Bottle trap with 76mm seal						
32mm	nr	0.25	2.40	1.91	0.65	4.96
40mm	nr	0.30	2.88	2.52	0.81	6.21
Tubular P trap						
32mm	nr	0.27	2.59	1.76	0.65	5.00
40mm	nr	0.32	3.07	2.06	0.77	5.90
Tubular S trap						
32mm	nr	0.27	2.59	2.49	0.76	5.84
40mm	nr	0.32	3.07	2.91	0.90	6.88
Adjustable tubular P trap						
32mm	nr	0.30	2.88	2.21	0.76	5.85
40mm	nr	0.35	3.36	2.33	0.85	6.54
Running P trap						
40mm	nr	0.32	3.07	4.71	1.17	8.95

	Unit	Labour hours	Net labour (£)	Net material (£)	O'heads /profit (£)	Total (£)

SOIL PIPES

uPVC, Terrain soil system 100, solvent-welded joints, to brickwork with holderbats at 1250mm maximum centres, plugging

	Unit	Labour hours	Net labour (£)	Net material (£)	O'heads /profit (£)	Total (£)
82mm diameter pipe	m	0.33	3.17	8.39	1.73	13.29
Extra over for						
bend, 92.5 degrees	nr	0.30	2.88	7.67	1.58	12.13
bend, 135 degrees	nr	0.30	2.88	7.67	1.58	12.13
branch, single, 92.5 degrees	nr	0.35	3.36	10.66	0.35	14.37
branch, single, 135 degrees	nr	0.35	3.36	10.66	0.35	14.37
vent cowl	nr	0.20	1.92	2.14	0.61	4.67
weathered apron	nr	0.22	2.11	2.14	0.64	4.89
110mm diameter pipe	m	0.37	3.55	9.75	2.00	15.30
Extra over for						
bend, 92.5 degrees	nr	0.34	3.26	10.49	2.06	15.81
bend, 135 degrees	nr	0.34	3.26	10.49	2.06	15.81
bend, variable, spigot/spigot	nr	0.34	3.26	14.94	2.73	20.93
bend, variable, spigot/socket	nr	0.34	3.26	14.94	2.73	20.93
bend, WC connecting, 92.5 degrees	nr	0.34	3.26	11.17	2.16	16.59
bend, access, 92.5 degrees	nr	0.34	3.26	23.54	4.02	30.82
branch, single, 92.5 degrees	nr	0.39	3.74	13.72	2.62	20.08
branch, single, 104 degrees	nr	0.39	3.74	13.88	2.64	20.26
branch, single, 135 degrees	nr	0.39	3.74	13.88	2.64	20.26
branch, single, access	nr	0.39	3.74	26.88	4.59	35.21
branch, double, 92.5 degrees	nr	0.39	3.74	32.36	5.42	41.52
branch, double, 104 degrees	nr	0.39	3.74	32.36	5.42	41.52
branch, double, 135 degrees	nr	0.39	3.74	33.83	5.64	43.21
branch, corner, 92.5 degrees	nr	0.39	3.74	58.55	9.34	71.63
bossed pipe connector, solvent welded sockets, single	nr	0.34	3.26	3.34	0.99	7.59
bossed pipe connector, spigot boss and solvent welded sockets	nr	0.34	3.26	3.47	1.01	7.74
connector, WC horizontal outlet	nr	0.30	2.88	9.48	1.85	14.21
connector, WC variable	nr	0.30	2.88	15.52	2.76	21.16

RATES FOR MEASURED WORK

Terrain soil system 100 (cont'd)	Unit	Labour hours	Net labour (£)	Net material (£)	O'heads /profit (£)	Total (£)
vent cowl	nr	0.22	2.11	2.16	0.64	4.91
weathered apron	nr	0.24	2.30	2.52	0.72	5.54
160mm diameter pipe	m	0.42	4.03	18.77	3.42	26.22
Extra over for						
bend, 92.5 degrees	nr	0.37	3.55	13.03	2.49	19.07
bend, 135 degrees	nr	0.37	3.55	31.91	5.32	40.78
branch, single, 92.5 degrees	nr	0.48	4.61	72.26	11.53	88.40
branch, single, 135 degrees	nr	0.48	4.61	46.42	7.65	58.68
branch, double, 92.5 degrees	nr	0.48	4.61	50.54	8.27	63.42
branch, double, 104 degrees	nr	0.48	4.61	56.39	9.15	70.15
vent cowl	nr	0.24	2.30	5.66	1.19	9.15
weathered apron	nr	0.26	2.50	7.61	1.52	11.63

**uPVC, Marley soil system, spigot
and socket seal joints, to
brickwork with brackets at 1250mm
maximum centres, plugging**

	Unit	Labour hours	Net labour (£)	Net material (£)	O'heads /profit (£)	Total (£)
82mm diameter pipe	m	0.30	2.88	8.60	1.72	13.20
Extra over for						
bend, spigot/socket	nr	0.27	2.59	7.56	1.52	11.67
bend, solvent socket	nr	0.27	2.59	7.56	1.52	11.67
bend, adjustable	nr	0.27	2.59	12.66	2.29	17.54
offset bend, socket	nr	0.27	2.59	7.56	1.52	11.67
offset bend, spigot	nr	0.27	2.59	6.38	1.35	10.32
branch, socket/socket/solvent socket, plain	nr	0.31	2.98	11.08	2.11	16.17
branch, adjustable socket/ spigot, 3 boss upstand	nr	0.31	2.98	11.85	2.22	17.05
boss branch, 3 boss upstand	nr	0.31	2.98	5.62	1.29	9.89
boss branch, single 50mm, 3 boss upstand/solvent sockets	nr	0.31	2.98	6.00	1.35	10.33
rear access bend, adjustable	nr	0.27	2.59	19.90	3.37	25.86
rear access branch, adjustable	nr	0.31	2.98	20.93	3.59	27.50
vent terminal	nr	0.20	1.92	2.04	0.59	4.55
vent slate cap	nr	0.20	1.92	2.75	0.70	5.37
WC connector, straight	nr	0.27	2.59	3.30	0.88	6.77

FOUL DRAINS ABOVE GROUND

	Unit	Labour hours	Net labour (£)	Net material (£)	O'heads /profit (£)	Total (£)
110mm diameter pipe	m	0.34	3.26	8.05	1.70	13.01
Extra over for						
bend, spigot/socket	nr	0.31	2.98	9.99	1.95	14.92
bend, solvent socket	nr	0.31	2.98	9.99	1.95	14.92
bend, adjustable	nr	0.31	2.98	12.66	2.35	17.99
offset bend, end socket	nr	0.31	2.98	8.40	1.71	13.09
branch, single socket/spigot, plain	nr	0.36	3.46	15.38	2.83	21.67
branch, single socket/spigot, 5 boss upstand	nr	0.36	3.46	13.00	2.47	18.93
branch, socket/socket/solvent socket, 3 boss upstand	nr	0.36	3.46	15.38	2.83	21.67
branch, socket/socket/solvent socket, plain	nr	0.36	3.46	15.38	2.83	21.67
branch, double socket/spigot, corner	nr	0.36	3.46	53.52	8.55	65.53
branch, double socket/spigot, 2 boss upstand/2 solvent sockets	nr	0.36	3.46	29.68	4.97	38.11
multi-branch	nr	0.36	3.46	26.34	4.47	34.27
multi-branch, with socket	nr	0.36	3.46	4.55	1.20	9.21
multi-branch, with access cap	nr	0.36	3.46	8.14	1.74	13.34
boss branch, collar boss	nr	0.36	3.46	15.37	2.82	21.65
boss branch, single 40mm	nr	0.36	3.46	6.91	1.56	11.93
boss branch, single 32mm	nr	0.36	3.46	6.91	1.56	11.93
boss branch, single 50mm, all solvent connections	nr	0.36	3.46	3.23	1.00	7.69
boss branch, single 40mm, double solvent sockets	nr	0.36	3.46	3.00	0.97	7.43
boss branch, single 32mm, double solvent sockets	nr	0.36	3.46	3.00	0.97	7.43
boss branch, 4 boss upstands	nr	0.36	3.46	6.56	1.50	11.52
boss branch, single 50mm, 3 boss upstand/solvent sockets	nr	0.36	3.46	6.00	1.42	10.88
rear access bend, plain	nr	0.31	2.98	23.27	3.94	30.19
rear access branch, 2 boss upstand	nr	0.36	3.46	27.39	4.63	35.48
roof cowl	nr	0.20	1.92	11.68	2.04	15.64
vent terminal	nr	0.20	1.92	2.26	0.63	4.81
weathering collar	nr	0.25	2.40	2.26	0.70	5.36

Marley soil system (cont'd)	Unit	Labour hours	Net labour (£)	Net material (£)	O'heads /profit (£)	Total (£)
WC connector, straight	nr	0.31	2.98	3.51	0.97	7.46
adaptor, to salt glazed socket	nr	0.25	2.40	5.30	1.16	8.86
adaptor, to cast iron socket	nr	0.25	2.40	12.86	2.29	17.55
160mm diameter pipe	m	0.37	3.55	18.24	3.27	25.06
Extra over for						
bend, spigot/socket	nr	0.34	3.26	23.86	4.07	31.19
bend, solvent socket	nr	0.34	3.26	23.86	4.07	31.19
bend, adjustable	nr	0.34	3.26	25.45	4.31	33.02
offset bend, socket	nr	0.34	3.26	24.55	4.17	31.98
offset bend, spigot	nr	0.34	3.26	22.50	3.86	29.62
branch, single socket/spigot, plain	nr	0.37	3.55	39.30	6.43	49.28
branch, triple solvent socket, 3 boss upstand	nr	0.37	3.55	28.69	4.84	37.08
branch, socket/socket/solvent socket, plain	nr	0.37	3.55	31.35	5.24	40.14
rear access bend, adjustable	nr	0.34	3.26	36.60	5.98	45.84
vent terminal	nr	0.20	1.92	5.95	1.18	9.05
vent slate cap	nr	0.20	1.92	10.48	1.86	14.26
adaptor, to cast iron socket	nr	0.25	2.40	34.98	5.61	42.99

Cast iron to BS416, Drainage System Ltd 'Timesaver' system flexible joints, to brickwork with holderbats at 2000mm maximum centres, plugging

	Unit	Labour hours	Net labour (£)	Net material (£)	O'heads /profit (£)	Total (£)
100mm diameter pipe	m	0.55	5.28	19.93	3.78	28.99
Extra over for						
bend, short radius	nr	0.45	4.32	13.92	2.74	20.98
bend, short radius, access	nr	0.45	4.32	29.45	5.07	38.84
bend, long radius	nr	0.45	4.32	22.56	4.03	30.91
bend, long radius, access	nr	0.45	4.32	38.08	6.36	48.76
bend, long radius, rest	nr	0.45	4.32	27.34	4.75	36.41
branch, 87.5 degrees	nr	0.50	4.80	21.56	3.95	30.31
branch, 76 degrees	nr	0.50	4.80	21.56	3.95	30.31
branch, 67.5 degrees	nr	0.50	4.80	21.56	3.95	30.31
branch, 45 degrees	nr	0.50	4.80	21.56	3.95	30.31
branch, access, 87.5 degrees	nr	0.50	4.80	37.08	6.28	48.16
branch, access, 76 degrees	nr	0.50	4.80	37.08	6.28	48.16

FOUL DRAINS ABOVE GROUND

	Unit	Labour hours	Net labour (£)	Net material (£)	O'heads /profit (£)	Total (£)
branch, access, 67.5 degrees	nr	0.50	4.80	37.08	6.28	48.16
branch, access, 45 degrees	nr	0.50	4.80	37.08	6.28	48.16
branch, double, 87.5 degrees	nr	0.50	4.80	26.65	4.72	36.17
branch, double, 76 degrees	nr	0.50	4.80	26.65	4.72	36.17
branch, double, 67.5 degrees	nr	0.50	4.80	26.65	4.72	36.17
branch, double, 45 degrees	nr	0.50	4.80	26.65	4.72	36.17
branch, double access, 87.5 degrees	nr	0.50	4.80	42.19	7.05	54.04
branch, double access, 76 degrees	nr	0.50	4.80	42.19	7.05	54.04
branch, double access, 67.5 degrees	nr	0.50	4.80	42.19	7.05	54.04
branch, double access, 45 degrees	nr	0.50	4.80	42.19	7.05	54.04
branch, parallel	nr	0.50	4.80	22.56	4.10	31.46
change piece, to soil	nr	0.40	3.84	10.51	2.15	16.50
change piece, to drain	nr	0.40	3.84	27.34	4.68	35.86
change piece, to clay	nr	0.40	3.84	19.31	3.47	26.62
WC connector	nr	0.45	4.32	15.16	2.92	22.40
WC connecting bend, 300mm tail	nr	0.45	4.32	24.34	4.30	32.96
long tail bend, 87.5 degrees	nr	0.45	4.32	42.19	6.98	53.49
long tail bend, 76 degrees	nr	0.45	4.32	51.88	8.43	64.63
offset, 75mm projection	nr	0.45	4.32	14.64	2.84	21.80
offset, 115mm projection	nr	0.45	4.32	17.63	3.29	25.24
offset, 150mm projection	nr	0.45	4.32	17.47	3.27	25.06
offset, 225mm projection	nr	0.45	4.32	17.47	3.27	25.06
offset, 300mm projection	nr	0.45	4.32	22.56	4.03	30.91
roof vent connector for asphalt	nr	0.35	3.36	30.70	5.11	39.17
roof vent connector for felt	nr	0.35	3.36	76.23	11.94	91.53
roof outlet	nr	0.35	3.36	43.45	7.02	53.83

	Unit	Labour hours	Net labour (£)	Net material (£)	O'heads /profit (£)	Total (£)

R12 DRAINAGE BELOW GROUND

Labour rate 8.00 per hour

HAND EXCAVATION

Excavate trench for drain, support sides, grade and ram bottom, backfill and consolidate with excavated material and remove surplus to skip for pipes not exceeding 200mm diameter, average depth of trench

	Unit	Labour hours	Net labour (£)	Net material (£)	O'heads /profit (£)	Total (£)
0.50m	m	1.20	9.60	0.00	1.44	11.04
0.75m	m	1.95	15.60	0.00	2.34	17.94
1.00m	m	2.58	20.64	0.00	3.10	23.74
1.25m	m	4.05	32.40	0.00	4.86	37.26
1.50m	m	5.10	40.80	0.00	6.12	46.92
1.75m	m	5.90	47.20	0.00	7.08	54.28
2.00m	m	6.30	50.40	0.00	7.56	57.96
2.25m	m	8.25	66.00	0.00	9.90	75.90
2.50m	m	9.45	75.60	0.00	11.34	86.94
2.75m	m	10.75	86.00	0.00	12.90	98.90
3.00m	m	12.00	96.00	0.00	14.40	110.40

Excavate trench for drain, support sides, grade and ram bottom, backfill and consolidate with excavated material and remove surplus to skip for pipes 225mm diameter, average depth of trench

	Unit	Labour hours	Net labour (£)	Net material (£)	O'heads /profit (£)	Total (£)
0.50m	m	1.32	10.56	0.00	1.58	12.14
0.75m	m	2.10	16.80	0.00	2.52	19.32
1.00m	m	2.84	22.72	0.00	3.41	26.13
1.25m	m	4.45	35.60	0.00	5.34	40.94
1.50m	m	5.60	44.80	0.00	6.72	51.52
1.75m	m	6.50	52.00	0.00	7.80	59.80
2.00m	m	6.95	55.60	0.00	8.34	63.94
2.25m	m	9.00	72.00	0.00	10.80	82.80

DRAINAGE BELOW GROUND

	Unit	Labour hours	Net labour (£)	Net material (£)	O'heads /profit (£)	Total (£)
2.50m	m	10. 40	83. 20	0. 00	12. 48	95. 68
2.75m	m	11. 85	94. 80	0. 00	14. 22	109. 02
3.00m	m	13. 20	105. 60	0. 00	15. 84	121. 44

Excavate trench for drain,
support sides, grade and ram
bottom, backfill and consolidate
with excavated material and
remove surplus to skip for pipes
300mm diameter, average depth of
trench

0.50m	m	1. 50	12. 00	0. 00	1. 80	13. 80
0.75m	m	2. 30	18. 40	0. 00	2. 76	21. 16
1.00m	m	3. 05	24. 40	0. 00	3. 66	28. 06
1.25m	m	4. 90	39. 20	0. 00	5. 88	45. 08
1.50m	m	6. 15	49. 20	0. 00	7. 38	56. 58
1.75m	m	7. 15	57. 20	0. 00	8. 58	65. 78
2.00m	m	7. 65	61. 20	0. 00	9. 18	70. 38
2.25m	m	9. 90	79. 20	0. 00	11. 88	91. 08
2.50m	m	11. 50	92. 00	0. 00	13. 80	105. 80
2.75m	m	13. 05	104. 40	0. 00	15. 66	120. 06
3.00m	m	14. 50	116. 00	0. 00	17. 40	133. 40

Extra for breaking up

concrete 100mm thick	m3	0. 90	7. 20	0. 00	1. 08	8. 28
tarmacadam 75mm thick	m3	0. 50	4. 00	0. 00	0. 60	4. 60
hardcore 100mm thick	m3	0. 60	4. 80	0. 00	0. 72	5. 52
plain concrete	m3	7. 00	56. 00	0. 00	8. 40	64. 40
reinforced concrete	m3	8. 00	64. 00	0. 00	9. 60	73. 60
soft rock	m3	10. 00	80. 00	0. 00	12. 00	92. 00
hard rock	m3	11. 00	88. 00	0. 00	13. 20	101. 20

RATES FOR MEASURED WORK

	Unit	Labour hours	Net labour (£)	Net material (£)	Net plant (£)	O'heads /profit (£)	Total (£)

MACHINE EXCAVATION

Excavate trench for drain,
support sides, grade and ram
bottom, backfill and consolidate
with excavated material and
remove surplus to skip for pipes
not exceeding 200mm diameter,
average depth of trench

	Unit	Labour hours	Net labour (£)	Net material (£)	Net plant (£)	O'heads /profit (£)	Total (£)
0.50m	m	0.80	6.40	0.00	0.51	1.04	7.95
0.75m	m	1.20	9.60	0.00	0.69	1.54	11.83
1.00m	m	1.60	12.80	0.00	0.86	2.05	15.71
1.25m	m	1.80	14.40	0.00	1.04	2.32	17.76
1.50m	m	2.00	16.00	0.00	1.27	2.59	19.86
1.75m	m	2.40	19.20	0.00	1.43	3.09	23.72
2.00m	m	2.80	22.40	0.00	1.62	3.60	27.62
2.25m	m	3.10	24.80	0.00	1.78	3.99	30.57
2.50m	m	3.40	27.20	0.00	1.96	4.37	33.53
2.75m	m	3.70	29.60	0.00	2.13	4.76	36.49
3.00m	m	4.00	32.00	0.00	2.31	5.15	39.46

Excavate trench for drain,
support sides, grade and ram
bottom, backfill and consolidate
with excavated material and
remove surplus to skip for pipes
225mm diameter, average depth of
trench

	Unit	Labour hours	Net labour (£)	Net material (£)	Net plant (£)	O'heads /profit (£)	Total (£)
0.50m	m	0.40	3.20	0.00	0.57	0.57	4.34
0.75m	m	1.30	10.40	0.00	0.74	1.67	12.81
1.00m	m	1.75	14.00	0.00	0.92	2.24	17.16
1.25m	m	1.95	15.60	0.00	1.09	2.50	19.19
1.50m	m	2.20	17.60	0.00	1.32	2.84	21.76
1.75m	m	2.60	20.80	0.00	1.50	3.35	25.65
2.00m	m	3.00	24.00	0.00	1.67	3.85	29.52
2.25m	m	3.30	26.40	0.00	1.91	4.25	32.56

DRAINAGE BELOW GROUND

	Unit	Labour hours	Net labour (£)	Net material (£)	Net plant (£)	O'heads /profit (£)	Total (£)
2.50m	m	3.60	28.80	0.00	2.08	4.63	35.51
2.75m	m	4.00	32.00	0.00	2.25	5.14	39.39
3.00m	m	4.40	35.20	0.00	2.48	5.65	43.33

Excavate trench for drain,
support sides, grade and ram
bottom, backfill and consolidate
with excavated material and
remove surplus to skip for pipes
300mm diameter, average depth of
trench

	Unit	Labour hours	Net labour (£)	Net material (£)	Net plant (£)	O'heads /profit (£)	Total (£)
0.50m	m	1.00	8.00	0.00	0.63	1.29	9.92
0.75m	m	1.45	11.60	0.00	0.81	1.86	14.27
1.00m	m	1.90	15.20	0.00	0.98	2.43	18.61
1.25m	m	2.15	17.20	0.00	1.27	2.77	21.24
1.50m	m	2.40	19.20	0.00	1.43	3.09	23.72
1.75m	m	2.85	22.80	0.00	1.60	3.66	28.06
2.00m	m	3.30	26.40	0.00	1.96	4.25	32.61
2.25m	m	3.65	29.20	0.00	2.14	4.70	36.04
2.50m	m	4.00	32.00	0.00	2.31	5.15	39.46
2.75m	m	4.40	35.20	0.00	2.54	5.66	43.40
3.00m	m	4.90	39.20	0.00	2.70	6.28	48.18

Extra for breaking up

	Unit	Labour hours	Net labour (£)	Net material (£)	Net plant (£)	O'heads /profit (£)	Total (£)
concrete 100mm thick	m3	0.40	3.20	0.00	1.84	0.76	5.80
tarmacadam 75mm thick	m3	0.24	1.92	0.00	1.15	0.46	3.53
hardcore 100mm thick	m3	0.28	2.24	0.00	1.27	0.53	4.04
plain concrete	m3	2.30	18.40	0.00	17.32	5.36	41.08
reinforced concrete	m3	3.50	28.00	0.00	20.47	7.27	55.74
soft rock	m3	4.00	32.00	0.00	11.55	6.53	50.08
hard rock	m3	4.00	32.00	0.00	20.47	7.87	60.34

BEDS AND COVERINGS

Sand bed in trench under 100mm
diameter pipe, thickness

	Unit	Labour hours	Net labour (£)	Net material (£)	Net plant (£)	O'heads /profit (£)	Total (£)
100mm	m	0.10	0.70	0.85	0.00	0.23	1.78
150mm	m	0.12	0.84	1.27	0.00	0.32	2.43

RATES FOR MEASURED WORK

Beds and coverings (cont'd)	Unit	Labour hours	Net labour (£)	Net material (£)	O'heads /profit (£)	Total (£)
Sand bed in trench under 150mm diameter pipe, thickness						
100mm	m	0.11	0.77	1.12	0.28	2.17
150mm	m	0.13	0.91	1.68	0.39	2.98
Sand bed in trench under 225mm diameter pipe, thickness						
100mm	m	0.14	0.98	1.35	0.35	2.68
150mm	m	0.16	1.12	2.03	0.47	3.62
Granular filling to bed in trench under 100mm diameter pipe, thickness						
100mm	m	0.12	0.84	0.67	0.23	1.74
150mm	m	0.14	0.98	1.01	0.30	2.29
Granular filling to bed in trench under 150mm diameter pipe, thickness						
100mm	m	0.13	0.91	0.81	0.26	1.98
150mm	m	0.15	1.05	1.22	0.34	2.61
Granular filling to bed in trench under 225mm diameter pipe, thickness						
100mm	m	0.16	1.12	1.01	0.32	2.45
150mm	m	0.18	1.26	1.52	0.42	3.20
Concrete Mix A bed in trench under 100mm diameter pipe, thickness						
100mm	m	0.24	1.68	3.06	0.71	5.45
150mm	m	0.28	1.96	4.59	0.98	7.53

DRAINAGE BELOW GROUND

	Unit	Labour hours	Net labour (£)	Net material (£)	O'heads /profit (£)	Total (£)
Concrete Mix A bed in trench under 150mm diameter pipe, thickness						
100mm	m	0.26	1.82	3.57	0.81	6.20
150mm	m	0.30	2.10	5.36	1.12	8.58
Concrete Mix A bed in trench under 225mm diameter pipe, thickness						
100mm	m	0.32	2.24	4.08	0.95	7.27
150mm	m	0.36	2.52	6.12	1.30	9.94
Granular filling in bed and haunching to 100mm diameter pipe, bed thickness						
100mm	m	0.24	1.68	1.34	0.45	3.47
150mm	m	0.28	1.96	2.01	0.60	4.57
Granular filling in bed and haunching to 150mm diameter pipe, bed thickness						
100mm	m	0.26	1.82	2.01	0.57	4.40
150mm	m	0.30	2.10	3.02	0.77	5.89
Granular filling in bed and haunching to 225mm diameter pipe, bed thickness						
100mm	m	0.32	2.24	3.22	0.82	6.28
150mm	m	0.36	2.52	4.83	1.10	8.45
Concrete Mix A in bed and haunching to 100mm diameter pipe, bed thickness						
100mm	m	0.48	3.36	6.12	1.42	10.90
150mm	m	0.56	3.92	9.18	1.96	15.06

RATES FOR MEASURED WORK

Beds and coverings (cont'd)	Unit	Labour hours	Net labour (£)	Net material (£)	O'heads /profit (£)	Total (£)
Concrete Mix A in bed and haunching to 150mm diameter pipe, bed thickness						
100mm	m	0.52	3.64	8.93	1.89	14.46
150mm	m	0.60	4.20	13.40	2.64	20.24
Concrete Mix A in bed and haunching to 225mm diameter pipe, bed thickness						
100mm	m	0.60	4.20	12.24	2.47	18.91
150mm	m	0.65	4.55	18.36	3.44	26.35
Granular filling in bed and surround to 100mm diameter pipe, bed thickness						
100mm	m	0.36	2.52	2.35	0.73	5.60
150mm	m	0.42	2.94	3.53	0.97	7.44
Granular filling in bed and surround to 150mm diameter pipe, bed thickness						
100mm	m	0.39	2.73	3.22	0.89	6.84
150mm	m	0.45	3.15	4.83	1.20	9.18
Granular filling in bed and surround to 225mm diameter pipe, bed thickness						
100mm	m	0.48	3.36	4.70	1.21	9.27
150mm	m	0.54	3.78	5.03	1.32	10.13
Concrete Mix A in bed and surround to 100mm diameter pipe, bed thickness						
100mm	m	0.72	5.04	10.71	2.36	18.11
150mm	m	0.84	5.88	16.08	3.29	25.25

DRAINAGE BELOW GROUND

	Unit	Labour hours	Net labour (£)	Net material (£)	O'heads /profit (£)	Total (£)
Concrete Mix A in bed and surround to 150mm diameter pipe, bed thickness						
100mm	m	0. 78	5. 46	12. 24	2. 65	20. 35
150mm	m	1. 20	8. 40	18. 76	4. 07	31. 23
Concrete Mix A in bed and surround to 225mm diameter pipe, bed thickness						
100mm	m	0. 90	6. 30	14. 28	3. 09	23. 67
150mm	m	0. 98	6. 86	21. 42	4. 24	32. 52

PIPEWORK

	Unit	Labour hours	Net labour (£)	Net material (£)	O'heads /profit (£)	Total (£)
Hepworths Supersleve vitrified clay drain pipes, spigot and socket joints with sealing rings 100mm diameter						
laid in trenches	m	0. 36	2. 88	2. 33	0. 78	5. 99
in lengths not exceeding 3m	m	0. 54	4. 32	2. 33	1. 00	7. 65
coupling	nr	0. 05	0. 40	1. 28	0. 25	1. 93
bends	nr	0. 30	2. 40	2. 68	0. 76	5. 84
rest bend	nr	0. 30	2. 40	4. 48	1. 03	7. 91
single junction	nr	0. 30	2. 40	5. 64	1. 21	9. 25
Hepworth's Supersleve vitrified clay drain pipes, spigot and socket joints with sealing rings 150mm diameter						
laid in trenches	m	0. 40	3. 20	5. 35	1. 28	9. 83
in lengths not exceeding 3m	m	0. 60	4. 80	5. 35	1. 52	11. 67
coupling	nr	0. 07	0. 56	3. 14	0. 56	4. 26
bends	nr	0. 35	2. 80	6. 76	1. 43	10. 99
rest bend	nr	0. 35	2. 80	8. 95	1. 76	13. 51
single junction	nr	0. 35	2. 80	9. 33	1. 82	13. 95

Pipework (cont'd)	Unit	Labour hours	Net labour (£)	Net material (£)	O'heads /profit (£)	Total (£)
Hepworth's Hepseal vitrified clay drain pipes, spigot and socket joints with sealing rings 100mm diameter						
laid in trenches	m	0. 36	2. 88	6. 54	1. 41	10. 83
in lengths not exceeding 3m	m	0. 54	4. 32	6. 54	1. 63	12. 49
bends	nr	0. 30	2. 40	8. 81	1. 68	12. 89
rest bend	nr	0. 30	2. 40	11. 00	2. 01	15. 41
single junction	nr	0. 30	2. 40	12. 26	2. 20	16. 86
Hepworth's Hepseal vitrified clay drain pipes, spigot and socket joints with sealing rings 150mm diameter						
laid in trenches	m	0. 40	3. 20	8. 57	1. 77	13. 54
in lengths not exceeding 3m	m	0. 60	4. 80	8. 57	2. 01	15. 38
bends	nr	0. 35	2. 80	14. 55	2. 60	19. 95
rest bend	nr	0. 35	2. 80	17. 35	3. 02	23. 17
single junction	nr	0. 35	2. 80	19. 01	3. 27	25. 08
Hepworth's Hepseal vitrified clay drain pipes, spigot and socket joints with sealing rings 225mm diameter						
laid in trenches	m	0. 60	4. 80	15. 08	2. 98	22. 86
in lengths not exceeding 3m	m	0. 90	7. 20	15. 08	3. 34	25. 62
bends	nr	0. 45	3. 60	29. 82	5. 01	38. 43
rest bend	nr	0. 45	3. 60	41. 54	6. 77	51. 91
single junction	nr	0. 45	3. 60	44. 83	7. 26	55. 69

GULLIES

	Unit	Labour hours	Net labour (£)	Net material (£)	O'heads /profit (£)	Total (£)
Vitrified clay gully with 100mm diameter outlet, 150mm square gully grid jointed to drain, surrounded in concrete	nr	1. 50	12. 00	24. 81	5. 52	42. 33
Yard gully, trapped, 150mm diameter with 100mm outlet, 200mm square gully grid, surrounded in concrete	nr	1. 50	12. 00	47. 53	1. 49	61. 02

DRAINAGE BELOW GROUND

	Unit	Labour hours	Net labour (£)	Net material (£)	O'heads /profit (£)	Total (£)
MANHOLES						
Excavate by hand for manhole not exceeding						
1m deep	m3	4.00	28.00	0.00	4.20	32.20
2m deep	m3	4.50	31.50	0.00	4.72	36.22
3m deep	m3	5.65	39.55	0.00	5.93	45.48
Excavate by machine for manhole not exceeding						
1m deep	m3	0.25	2.00	5.54	1.13	8.67
2m deep	m3	0.28	2.24	6.24	1.27	9.75
3m deep	m3	0.50	4.00	6.93	1.64	12.57
Earthwork support not exceeding 2m between opposing faces not exceeding						
1m deep	m2	0.40	2.80	1.05	0.58	4.43
2m deep	m2	0.45	3.15	1.15	0.64	4.94
3m deep	m2	0.55	3.85	1.26	0.77	5.88
Load excavated material into barrows, wheel average 50m and load into skip	m3	2.80	19.60	0.00	2.94	22.54
Site mixed concrete A in manhole bed, thickness						
100 to 150mm	m3	2.00	14.00	51.02	9.75	74.77
150 to 300mm	m3	1.70	11.90	51.02	9.44	72.36
Site mixed concrete B in benching to manholes average 225mm thick	m3	3.40	23.80	59.26	12.46	95.52
Common bricks in cement mortar walls to manholes one brick thick	m2	4.25	34.00	21.70	8.36	64.06
Class 'B' engineering bricks in cement mortar in walls to manholes one brick thick	m2	4.40	35.20	47.23	2.06	84.49

349

RATES FOR MEASURED WORK

Manholes (cont'd)	Unit	Labour hours	Net labour (£)	Net material (£)	O'heads /profit (£)	Total (£)
Extra for fair face and flush pointing	m2	0. 25	2. 00	0. 00	0. 30	2. 30
Build in ends of pipes to one brick wall and make good						
small pipe	nr	0. 15	1. 20	0. 00	0. 18	1. 38
large pipe	nr	0. 20	1. 60	0. 00	0. 24	1. 84
Galvanized step irons built into brickwork	nr	0. 25	2. 00	4. 89	1. 03	7. 92
Cast iron manhole covers, frame bedded in cement mortar						
grade C light duty, size 600 x 450mm	nr	2. 40	19. 20	31. 92	7. 67	58. 79
grade B medium duty, single seal, size 600 x 450mm	nr	2. 40	19. 20	101. 80	18. 15	139. 15
Best quality vitrified clay channels, bedded and jointed in cement mortar						
Half section straight main channel						
100mm diameter	m	1. 10	8. 80	2. 57	1. 71	11. 37
150mm diameter	m	1. 25	10. 00	9. 64	2. 95	19. 64
Half section tapered main channel						
150 to 100mm diameter	m	1. 25	10. 00	11. 48	3. 22	21. 48
Half section 90 deg channel bends						
100mm diameter	nr	1. 10	8. 80	2. 73	1. 73	11. 53
150mm diameter	nr	1. 25	10. 00	4. 54	2. 18	14. 54
Three quarter section branch channel bends						
100mm diameter	nr	1. 10	8. 80	5. 63	2. 16	14. 43
150mm diameter	nr	1. 25	10. 00	9. 24	2. 89	19. 24

S

PIPED SUPPLY SYSTEMS

S10 Hot and cold water

MATERIAL COSTS

	Unit	Material supply (£)	Material waste (%)	Total (£)

MATERIAL COSTS

The following allowances have been made for waste in this section

Pipework - 7.5%

Pipe fittings - 5%

Stopcocks - 2.5%

Valves - 5%

RATES FOR MEASURED WORK

	Unit	Labour hours	Net labour (£)	Net material (£)	O'heads /profit (£)	Total (£)
S10 HOT AND COLD WATER						
Labour rate 9.60 per hour						
Copper pipe to BS2871, Table X, lead free pre-soldered capillary joints and fittings to BS864, two piece clips at 1250mm maximum centres						
15mm diameter, to timber	m	0.20	1.92	1.49	0.34	3.41
15mm diameter, plugged and screwed	m	0.22	2.11	1.77	0.58	4.46
Extra over for						
made bend	nr	0.10	0.96	0.00	0.14	1.10
straight coupling	nr	0.18	1.73	0.21	0.29	2.23
reduced coupling, 15 x 8mm	nr	0.18	1.73	1.62	0.50	3.85
reduced coupling, 15 x 10mm	nr	0.18	1.73	1.22	0.44	3.39
reduced coupling, 15 x 12mm	nr	0.18	1.73	1.30	0.45	3.48
adaptor coupling, imperial to metric, 1/2in	nr	0.18	1.73	0.95	0.40	3.08
straight female connector, 15mm x 1/2in	nr	0.18	1.73	1.41	0.47	3.61
female reducing connector, 15mm x 3/8in	nr	0.18	1.73	2.34	0.61	4.68
female reducing connector, 15mm x 3/4in	nr	0.18	1.73	3.89	0.84	6.46
straight male connector, 15mm x 1/2in	nr	0.18	1.73	1.13	0.43	3.29
male reducing connector, 15mm x 3/4in	nr	0.18	1.73	2.76	0.67	5.16
male reducing connector, 15mm x 3/8in	nr	0.18	1.73	1.97	0.55	4.25
lead connector	nr	0.18	1.73	0.91	0.40	3.04
tank connector, 15mm x 1/2in	nr	0.28	2.69	2.91	0.84	6.44
tank connector, 15mm x 3/4in	nr	0.28	2.69	6.51	1.38	10.58
reducer, 15 x 8mm	nr	0.18	1.73	0.99	0.41	3.13
reducer, 15 x 10mm	nr	0.18	1.73	0.99	0.41	3.13
reducer, 15 x 12mm	nr	0.18	1.73	0.98	0.41	3.12
female adaptor, 15mm x 1/2in	nr	0.18	1.73	2.21	0.59	4.53
male adaptor, 15mm x 1/2in	nr	0.18	1.73	2.21	0.59	4.53
adaptor, imperial to metric, 1/2 x 15mm	nr	0.18	1.73	0.86	0.39	2.98

HOT AND COLD WATER

	Unit	Labour hours	Net labour (£)	Net material (£)	O'heads /profit (£)	Total (£)
street elbow	nr	0.18	1.73	1.32	0.46	3.51
male elbow, 15mm x 1/2in	nr	0.18	1.73	2.56	0.64	4.93
reducing male elbow, 15 x 8mm	nr	0.18	1.73	3.69	0.81	6.23
female elbow, 15mm x 1/2in	nr	0.18	1.73	2.40	0.62	4.75
reducing female elbow, 15mm x 1/4in	nr	0.18	1.73	3.61	0.80	6.14
backplate elbow, 15mm x 1/2in	nr	0.28	2.69	2.82	0.83	6.34
flanged bend, 15mm x 1/2in	nr	0.22	2.11	5.22	1.10	8.43
flanged bend, 15mm x 3/4in	nr	0.22	2.11	5.22	1.10	8.43
overflow bend, 15mm x 1/2in	nr	0.22	2.11	6.39	1.27	9.77
slow bend	nr	0.22	2.11	1.58	0.55	4.24
return bend	nr	0.22	2.11	5.63	1.16	8.90
obtuse elbow	nr	0.18	1.73	1.29	0.45	3.47
tee, reduced branch, largest branch 15mm	nr	0.24	2.30	2.97	0.79	6.06
tee, one end and one branch reduced, largest branch, 15mm	nr	0.24	2.30	3.31	0.84	6.45
tee, both ends reduced, largest end, 15mm	nr	0.24	2.30	4.12	0.96	7.38
corner tee	nr	0.24	2.30	5.78	1.21	9.29
sweep tee, 90 degrees	nr	0.24	2.30	3.81	0.92	7.03
offset tee, 15mm equal	nr	0.24	2.30	6.02	1.25	9.57
double sweep tee, 15mm	nr	0.24	2.30	5.24	1.13	8.67
cross equal tee, 15mm	nr	0.28	2.69	4.83	1.13	8.65
stop end, 15mm	nr	0.18	1.73	1.72	0.52	3.97
tap connector, 15mm x 3/4in	nr	0.22	2.11	2.46	0.69	5.26
bent union adaptor, 15mm x 3/4in	nr	0.22	2.11	2.86	0.75	5.72
bent male union connector, 15mm x 1/2in	nr	0.22	2.11	5.12	1.08	8.31
bent female union connector, 15mm x 1/2in	nr	0.22	2.11	5.12	1.08	8.31
22mm diameter, to timber	m	0.21	2.02	2.76	0.72	5.50
22mm diameter, plugged and screwed	m	0.23	2.21	3.02	0.78	6.01
Extra over for						
made bend	nr	0.15	1.44	0.00	0.22	1.66
straight coupling	nr	0.22	2.11	0.45	0.38	2.94
reduced coupling, 22 x 8mm	nr	0.22	2.11	1.92	0.60	4.63
reduced coupling, 22 x 10mm	nr	0.22	2.11	1.92	0.60	4.63
reduced coupling, 22 x 15mm	nr	0.22	2.11	1.09	0.48	3.68
straight female connector, 22mm x 3/4in	nr	0.22	2.11	2.49	0.69	5.29

Pipework (cont'd)

	Unit	Labour hours	Net labour (£)	Net material (£)	O'heads /profit (£)	Total (£)
female reducing connector, 22mm x 1/2in	nr	0.22	2.11	5.80	1.19	9.10
female reducing connector, 22mm x 1in	nr	0.22	2.11	5.12	1.08	8.31
straight male connector, 22mm x 3/4in	nr	0.22	2.11	5.01	1.07	8.19
male reducing connector, 22mm x 1/2in	nr	0.22	2.11	3.50	0.84	6.45
male reducing connector, 22mm x 1in	nr	0.22	2.11	3.45	0.83	6.39
lead connector	nr	0.22	2.11	1.41	0.53	4.05
tank connector, 22mm x 1in	nr	0.30	2.88	10.40	1.99	15.27
reducer, 22 x 12mm	nr	0.22	2.11	1.75	0.58	4.44
female adaptor, 22mm x 3/4in	nr	0.22	2.11	3.29	0.81	6.21
male adaptor, 22mm x 3/4in	nr	0.22	2.11	3.29	0.81	6.21
adaptor, imperial to metric, 22mm x 3/4in	nr	0.22	2.11	0.96	0.46	3.53
reducing elbow, 22 x 15mm	nr	0.22	2.11	4.35	0.97	7.43
street elbow	nr	0.22	2.11	2.03	0.62	4.76
male elbow, 22mm x 3/4in	nr	0.22	2.11	3.64	0.86	6.61
female elbow, 22mm x 3/4in	nr	0.22	2.11	3.31	0.81	6.23
reducing female elbow, 22mm x 1/2in	nr	0.22	2.11	3.52	0.84	6.47
backplate elbow, 22mm x 3/4in	nr	0.00	0.00	6.17	0.93	7.10
flanged bend, 22mm x 3/4in	nr	0.24	2.30	5.30	1.14	8.74
overflow bend, 22mm x 3/4in	nr	0.24	2.30	6.65	1.34	10.29
slow bend	nr	0.24	2.30	2.69	0.75	5.74
return bend	nr	0.24	2.30	6.65	1.34	10.29
obtuse elbow	nr	0.24	2.30	2.12	0.66	5.08
tee, reduced branch, largest branch 22mm	nr	0.28	2.69	3.91	0.99	7.59
tee, both ends reduced, largest end, 22mm	nr	0.28	2.69	4.84	1.13	8.66
corner tee	nr	0.28	2.69	7.42	1.52	11.63
sweep tee, 90 degrees	nr	0.28	2.69	4.76	1.12	8.57
sweep tee, reduced branch largest branch, 22mm	nr	0.28	2.69	5.01	1.16	8.86
offset tee, 22mm equal	nr	0.28	2.69	6.60	1.39	10.68
double sweep tee, 22mm	nr	0.28	2.69	7.14	1.47	11.30
cross equal tee, 22mm	nr	0.32	3.07	7.76	1.62	12.45
stop end, 22mm	nr	0.20	1.92	2.61	0.68	5.21
tap connector, 22mm x 1/2in	nr	0.24	2.30	5.06	1.10	8.46
bent tap connector, 22mm x 1/2in	nr	0.24	2.30	6.12	1.26	9.68
bent tap connector, 22mm x 3/4in	nr	0.24	2.30	3.15	0.82	6.27

HOT AND COLD WATER

	Unit	Labour hours	Net labour (£)	Net material (£)	O'heads /profit (£)	Total (£)
bent union adaptor, 22mm x 1in	nr	0.24	2.30	4.15	0.97	7.42
bent male union connector, 22mm x 3/4in	nr	0.24	2.30	6.99	1.39	10.68
bent female union connector, 22mm x 3/4in	nr	0.24	2.30	6.99	1.39	10.68
28mm diameter, to timber	m	0.23	2.21	3.79	0.90	6.90
28mm diameter, plugged and screwed	m	0.25	2.40	4.04	0.97	7.41
Extra over for						
made bend	nr	0.20	1.92	0.00	0.29	2.21
straight coupling	nr	0.24	2.30	0.79	0.46	3.55
reduced coupling, 28 x 15mm	nr	0.24	2.30	2.55	0.73	5.58
reduced coupling, 28 x 22mm	nr	0.24	2.30	1.55	0.58	4.43
adaptor coupling, imperial to metric, 1 x 28mm	nr	0.24	2.30	1.73	0.60	4.63
flanged connector	nr	0.24	2.30	23.49	3.87	29.66
straight female connector, 28mm x 1in	nr	0.24	2.30	3.91	0.93	7.14
straight male connector, 28mm x 1in	nr	0.24	2.30	3.31	0.84	6.45
male reducing connector, 28mm x 3/4in	nr	0.24	2.30	5.00	1.09	8.39
lead connector	nr	0.24	2.30	2.02	0.65	4.97
tank connector 28mm x 1in	nr	0.32	3.07	5.92	1.35	10.34
tank connector, 28mm x 1 1/4in	nr	0.32	3.07	14.20	2.59	19.86
reducer, 28 x 15mm	nr	0.24	2.30	1.39	0.55	4.24
reducer, 28 x 22mm	nr	0.24	2.30	1.17	0.52	3.99
female adaptor, 28mm x 1in	nr	0.24	2.30	5.01	1.10	8.41
male adaptor, 28mm x 1in	nr	0.24	2.30	5.01	1.10	8.41
adaptor, imperial to metric, 1in x 28mm	nr	0.24	2.30	1.64	0.59	4.53
reducing elbow, 28 x 22mm	nr	0.24	2.30	6.78	1.36	10.44
street elbow	nr	0.24	2.30	3.69	0.90	6.89
male elbow, 28mm x 1in	nr	0.24	2.30	5.46	1.16	8.92
female elbow, 28mm x 1in	nr	0.24	2.30	5.18	1.12	8.60
slow bend	nr	0.26	2.50	4.47	1.05	8.02
return bend	nr	0.26	2.50	9.84	1.85	14.19
obtuse elbow	nr	0.26	2.50	3.45	0.89	6.84
equal tee, 28mm	nr	0.32	3.07	3.48	0.98	7.53
tee, reduced branch, largest branch 28mm	nr	0.32	3.07	3.74	1.02	7.83

RATES FOR MEASURED WORK

Pipework (cont'd)	Unit	Labour hours	Net labour (£)	Net material (£)	O'heads /profit (£)	Total (£)
tee, one end and one branch reduced, largest branch 28mm	nr	0.32	3.07	5.10	1.23	9.40
tee, both ends reduced, largest end 28mm	nr	0.32	3.07	4.84	1.19	9.10
sweep tee, 90 degrees	nr	0.32	3.07	9.86	1.94	14.87
sweep tee, reduced branch largest branch, 28mm	nr	0.32	3.07	8.48	1.73	13.28
double sweep tee	nr	0.32	3.07	10.85	2.09	16.01
cross equal tee	nr	0.32	3.07	11.12	2.13	16.32
stop end, 28mm	nr	0.22	2.11	2.95	0.76	5.82
bent union adaptor 28mm x 1 1/4in	nr	0.28	2.69	6.51	1.38	10.58
bent male union connector, 28mm x 1in	nr	0.28	2.69	10.21	1.94	14.84
bent female union connector, 28mm x 1in	nr	0.28	2.69	10.21	1.94	14.84
bent female union connector, 54mm x 2in	nr	0.42	4.03	10.21	2.14	16.38

HOT AND COLD WATER

	Unit	Labour hours	Net labour (£)	Net material (£)	O'heads /profit (£)	Total (£)
STOPCOCKS						
Stopcocks to BS1010, lead free pre-soldered capillary joints						
Gunmetal stopcock with brass headwork, copper x copper						
15mm	nr	0.20	1.92	3.15	0.76	5.83
22mm	nr	0.22	2.11	5.89	1.20	9.20
28mm	nr	0.28	2.69	16.74	2.91	22.34
Dezincification-resistant stopcock, copper x copper						
15mm	nr	0.20	1.92	7.92	1.48	11.32
22mm	nr	0.22	2.11	13.71	2.37	18.19
28mm	nr	0.28	2.69	22.84	3.83	29.36
VALVES						
Mistral radiator valve, plain finish, compression inlet x taper male union outlet						
8mm x 1/2in	nr	0.30	2.88	5.16	1.21	9.25
10mm x 1/2in	nr	0.30	2.88	5.16	1.21	9.25
15mm x 1/2in	nr	0.30	2.88	5.16	1.21	9.25
Mistral radiator valve, chromium-plated finish, compression inlet x taper male union outlet						
8mm x 1/2in	nr	0.30	2.88	5.51	1.26	9.65
10mm x 1/2in	nr	0.30	2.88	5.51	1.26	9.65
15mm x 1/2in	nr	0.30	2.88	5.51	1.26	9.65
Maxitwin radiator valve, plain finish, compression connection x taper male union outlet						
two 8mm x 1/2in	nr	0.30	2.88	12.59	2.32	17.79
two 10mm x 1/2in	nr	0.30	2.88	12.59	2.32	17.79

RATES FOR MEASURED WORK

	Unit	Labour hours	Net labour (£)	Net material (£)	O'heads /profit (£)	Total (£)
Maxitwin radiator valve, chromium-plated finish, compression connection x taper male union outlet						
two 8mm x 1/2in	nr	0.30	2.88	15.29	2.73	20.90
two 10mm x 1/2in	nr	0.30	2.88	16.79	2.95	22.62

COLD WATER STORAGE TANKS

Galvanized steel cistern, BS417
Part A with cover, reference

	Unit	Labour hours	Net labour (£)	Net material (£)	O'heads /profit (£)	Total (£)
SC10, 18 litres	nr	0.60	5.76	37.82	6.54	50.12
SC15, 36 litres	nr	0.65	6.24	42.85	7.36	56.45
SC20, 54 litres	nr	0.65	6.24	49.39	8.34	63.97
SC25, 68 litres	nr	0.70	6.72	51.28	8.70	66.70
SC30, 86 litres	nr	0.75	7.20	53.19	9.06	69.45
SC40, 114 litres	nr	0.80	7.68	60.49	10.23	78.40
SC50, 154 litres	nr	0.85	8.16	86.03	14.13	108.32
SC60, 191 litres	nr	0.85	8.16	91.35	14.93	114.44

Plastic cistern, BS4213 with lid, reference

	Unit	Labour hours	Net labour (£)	Net material (£)	O'heads /profit (£)	Total (£)
PC4, 18 litres	nr	0.60	5.76	7.21	1.95	14.92
PC15, 68 litres	nr	0.70	6.72	25.27	4.80	36.79
PC20, 91 litres	nr	0.75	7.20	26.91	5.12	39.23
PC25, 114 litres	nr	0.80	7.68	34.29	6.30	48.27
PC40, 182 litres	nr	0.90	8.64	61.62	10.54	80.80
PC50, 227 litres	nr	1.00	9.60	64.87	11.17	85.64

HOT WATER TANKS

Galvanized steel tank, BS417 with cover, reference

	Unit	Labour hours	Net labour (£)	Net material (£)	O'heads /profit (£)	Total (£)
T25/1, 36 litres	nr	0.65	6.24	123.18	19.41	148.83
T25/2, 36 litres	nr	0.65	6.24	125.09	19.70	151.03
T30/1, 86 litres	nr	0.75	7.20	131.96	20.87	160.03
T30/2, 86 litres	nr	0.75	7.20	133.59	21.12	161.91
T40, 114 litres	nr	0.80	7.68	161.79	25.42	194.89

HOT AND COLD WATER

	Unit	Labour hours	Net labour (£)	Net material (£)	O'heads /profit (£)	Total (£)
HOT WATER COPPER CYLINDERS						
Indirect cylinder to BS699 grade 3, reference						
ref 2, 96 litres	nr	0.50	4.80	118.79	18.54	142.13
ref 7, 117 litres	nr	0.55	5.28	124.85	19.52	149.65
ref 8, 140 litres	nr	0.65	6.24	204.33	31.59	242.16
ref 9, 162 litres	nr	0.70	6.72	180.54	28.09	215.35
INSULATION						
Preformed pipe lagging, fire-retardant foam, 19mm thick to pipe size						
15mm	nr	0.08	0.77	0.75	0.23	1.75
22mm	nr	0.08	0.77	0.92	0.25	1.94
28mm	nr	0.10	0.96	1.10	0.31	2.37
50mm thick glass-fibre-filled insulating jacket, fixing bands to tank size						
445 x 305 x 300mm	nr	0.35	3.36	3.12	0.97	7.45
495 x 368 x 362mm	nr	0.40	3.84	5.68	1.43	10.95
630 x 450 x 420mm	nr	0.45	4.32	6.95	1.69	12.96
665 x 490 x 515mm	nr	0.50	4.80	7.61	1.86	14.27
700 x 540 x 535mm	nr	0.60	5.76	10.75	2.48	18.99
995 x 605 x 595mm	nr	0.70	6.72	13.35	3.01	23.08
60mm thick glass-fibre-filled insulating jacket, fixing bands, to cylinder 450mm diameter, height						
300mm	nr	0.30	2.88	7.06	1.49	11.43
450mm	nr	0.35	3.36	7.61	1.65	12.62
600mm	nr	0.40	3.84	8.60	1.87	14.31

T

HEATING SYSTEMS

MATERIALS

	Unit	Material supply (£)	Material waste (%)	Total (£)

MATERIALS

The following allowances have been made for waste in this section

Boilers - 1%

Radiators - 1%

RATES FOR MEASURED WORK

	Unit	Labour hours	Net labour (£)	Net material (£)	O'heads /profit (£)	Total (£)
T10 GAS/OIL FIRED BOILERS						

Labour rate 9.60 per hour

Gas fired free standing boiler for domestic central heating and indirect hot water

Balanced flue, output

40,000 Btu	nr	4.80	46.08	579.34	93.81	719.23
50,000 Btu	nr	4.80	46.08	617.78	99.58	763.44
60,000 Btu	nr	4.80	46.08	661.66	106.16	813.90
80,000 Btu	nr	4.80	46.08	906.57	142.90	1095.55
100,000 Btu	nr	4.80	46.08	1135.80	177.28	1359.16

Open flue, output

40,000 Btu	nr	4.20	33.60	460.19	74.07	567.86
50,000 Btu	nr	4.20	40.32	485.07	78.81	604.20
60,000 Btu	nr	4.20	40.32	513.49	83.07	636.88
80,000 Btu	nr	4.20	40.32	657.64	104.69	802.65
100,000 Btu	nr	4.20	40.32	854.84	134.27	1029.43

Gas fired wall hung boiler for domestic central heating and indirect heating and indirect hot water

Balanced flue, output

30,000 Btu	nr	4.80	46.08	442.62	73.30	562.00
40,000 Btu	nr	4.80	46.08	500.10	81.93	628.11
50,000 Btu	nr	4.80	46.08	579.96	93.91	719.95
60,000 Btu	nr	4.80	46.08	705.24	112.70	864.02
80,000 Btu	nr	4.80	46.08	1161.05	181.07	1388.20

Open flue, output

35,000 Btu	nr	4.20	40.32	426.75	70.06	537.13
50,000 Btu	nr	4.20	40.32	472.70	76.95	589.97

	Unit	Labour hours	Net labour (£)	Net material (£)	O'heads /profit (£)	Total (£)
Oil fired boiler for domestic central heating and indirect heating and indirect hot water						
Balanced flue, output						
50,000 Btu	nr	4.80	46.08	945.92	148.80	1140.80
60,000 Btu	nr	4.80	46.08	979.82	153.88	1179.78
80,000 Btu	nr	4.80	46.08	1099.33	171.81	1317.22
100,000 Btu	nr	4.80	46.08	1230.24	191.45	1467.77
Conventional flue, output						
50,000 Btu	nr	4.20	33.60	960.24	149.08	1142.92
60,000 Btu	nr	4.20	40.32	991.15	154.72	1186.19
80,000 Btu	nr	4.20	40.32	1107.58	172.18	1320.08
100,000 Btu	nr	4.20	40.32	1219.88	189.03	1449.23
Fibre cement light quality flue pipe and fittings, joints in running lengths, pipe supports fixed to brickwork						
100mm	m	0.35	3.36	9.44	1.92	14.72
Extra over for						
bend	nr	0.40	3.84	5.98	1.47	11.29
bend with door	nr	0.40	3.84	10.46	2.15	16.45
equal tee	nr	0.45	4.32	2.37	1.00	7.69
cone caps	nr	0.40	3.84	12.88	2.51	19.23
GCJ terminal	nr	0.40	3.84	22.83	4.00	30.67
GLC terminal	nr	0.40	3.84	17.82	3.25	24.91
H piece	nr	0.45	4.32	28.73	4.96	38.01
blank cap	nr	0.40	3.84	3.45	1.09	8.38

RATES FOR MEASURED WORK

	Unit	Labour hours	Net labour (£)	Net material (£)	O'heads /profit (£)	Total (£)
T30 MEDIUM TEMPERATURE HOT WATER HEATING RADIATORS						
Steel panel radiators, Barlo, plugged and screwed to brickwork with concealed brackets, satin primer finish						
Single panel, 420mm high, length						
500mm	nr	1.00	9.60	15.72	3.80	29.12
650mm	nr	1.00	9.60	20.28	4.48	34.36
750mm	nr	1.05	10.08	22.28	4.85	37.21
950mm	nr	1.10	10.56	28.33	5.83	44.72
1200mm	nr	1.15	11.04	35.43	6.97	53.44
1450mm	nr	1.20	11.52	42.46	8.10	62.08
1550mm	nr	1.25	12.00	45.24	8.59	65.83
1700mm	nr	1.30	12.48	49.69	9.33	71.50
1850mm	nr	1.35	12.96	53.59	9.98	76.53
Single panel, 720mm high, length						
500mm	nr	1.10	10.56	26.63	5.58	42.77
650mm	nr	1.10	10.56	36.34	7.04	53.94
750mm	nr	1.15	11.04	39.44	7.57	58.05
850mm	nr	1.20	11.52	43.35	8.23	63.10
950mm	nr	1.20	11.52	48.26	8.97	68.75
1100mm	nr	1.25	12.00	55.60	10.14	77.74
1200mm	nr	1.25	12.00	60.51	10.88	83.39
1300mm	nr	1.30	12.48	65.30	11.67	89.45
1450mm	nr	1.30	12.48	72.54	12.75	97.77
1550mm	nr	1.35	12.96	77.33	13.54	103.83
1700mm	nr	1.40	13.44	73.32	13.01	99.77
1850mm	nr	1.45	13.92	64.96	11.83	90.71
Single Superplus convector, 520mm high, length						
500mm	nr	1.00	9.60	27.30	5.54	42.44
650mm	nr	1.00	9.60	34.98	6.69	51.27
750mm	nr	1.05	10.08	40.11	7.53	57.72
850mm	nr	1.10	10.56	45.13	8.35	64.04
950mm	nr	1.10	10.56	50.14	9.11	69.81
1100mm	nr	1.15	11.04	57.61	10.30	78.95

RADIATORS

	Unit	Labour hours	Net labour (£)	Net material (£)	O'heads /profit (£)	Total (£)
1200mm	nr	1.15	11.04	62.51	11.03	84.58
1300mm	nr	1.20	11.52	67.42	11.84	90.78
1450mm	nr	1.20	11.52	73.98	12.82	98.32
1550mm	nr	1.25	12.00	77.88	13.48	103.36
1700mm	nr	1.30	12.48	84.68	14.57	111.73
1850mm	nr	1.35	12.96	91.59	15.68	120.23
2000mm	nr	1.40	13.44	98.61	16.81	128.86

Single Superplus convector,
620mm high, length

	Unit	Labour hours	Net labour	Net material	O'heads /profit	Total
500mm	nr	1.10	10.56	30.30	6.13	46.99
650mm	nr	1.10	10.56	38.69	7.39	56.64
750mm	nr	1.15	11.04	44.46	8.32	63.82
850mm	nr	1.20	11.52	50.04	9.23	70.79
950mm	nr	1.20	11.52	55.60	10.07	77.19
1100mm	nr	1.25	12.00	63.84	11.38	87.22
1200mm	nr	1.25	12.00	69.30	12.19	93.49
1300mm	nr	1.30	12.48	73.87	12.95	99.30
1450mm	nr	1.30	12.48	81.78	14.14	108.40
1550mm	nr	1.35	12.96	87.12	15.01	115.09
1700mm	nr	1.40	13.44	94.71	16.22	124.37
1850mm	nr	1.45	13.92	102.84	17.51	134.27
2000mm	nr	1.50	14.40	110.65	18.76	143.81

Double Superplus convector,
520mm high, length

	Unit	Labour hours	Net labour	Net material	O'heads /profit	Total
500mm	nr	1.00	9.60	52.59	9.33	71.52
650mm	nr	1.00	9.60	68.51	11.72	89.83
750mm	nr	1.05	10.08	79.00	13.36	102.44
850mm	nr	1.10	10.56	86.02	14.49	111.07
950mm	nr	1.10	10.56	96.16	16.01	122.73
1100mm	nr	1.15	11.04	111.31	18.35	140.70
1200mm	nr	1.15	11.04	121.44	19.87	152.35
1300mm	nr	1.20	11.52	131.48	21.45	164.45
1450mm	nr	1.20	11.52	146.62	23.72	181.86
1550mm	nr	1.25	12.00	156.65	25.30	193.95
1700mm	nr	1.30	12.48	171.71	27.63	211.82
1850mm	nr	1.35	12.96	186.74	29.96	229.66
2000mm	nr	1.40	13.44	201.90	32.30	247.64

RATES FOR MEASURED WORK

	Unit	Labour hours	Net labour (£)	Net material (£)	O'heads /profit (£)	Total (£)
Double Superplus convector, 720mm high, length						
500mm	nr	1.00	9.60	69.20	11.82	90.62
650mm	nr	1.00	9.60	90.14	14.96	114.70
750mm	nr	1.05	10.08	104.06	17.12	131.26
850mm	nr	1.10	10.56	113.09	18.55	142.20
950mm	nr	1.10	10.56	126.24	20.52	157.32
1100mm	nr	1.15	11.04	146.30	23.60	180.94
1200mm	nr	1.15	11.04	159.55	25.59	196.18
1300mm	nr	1.20	11.52	172.81	27.65	211.98
1450mm	nr	1.20	11.52	192.65	30.63	234.80
1550mm	nr	1.25	12.00	205.90	32.69	250.59
1700mm	nr	1.30	12.48	225.74	35.73	273.95

Steel panel radiators, Potterton Myson, plugged and screwed to brickwork with concealed brackets, full finish in white

Supaline single panel, 440mm high, length

	Unit	Labour hours	Net labour (£)	Net material (£)	O'heads /profit (£)	Total (£)
480mm	nr	1.00	9.60	15.20	3.72	28.52
640mm	nr	1.00	9.60	19.52	4.37	33.49
800mm	nr	1.00	9.60	23.98	5.04	38.62
960mm	nr	1.10	10.56	28.26	5.82	44.64
1120mm	nr	1.10	10.56	32.44	6.45	49.45
1280mm	nr	1.15	11.04	36.72	7.16	54.92
1440mm	nr	1.20	11.52	40.91	7.86	60.29
1600mm	nr	1.25	12.00	45.09	8.56	65.65
1760mm	nr	1.30	12.48	49.17	9.25	70.90
1920mm	nr	1.35	12.96	53.45	9.96	76.37
2080mm	nr	1.40	13.44	57.43	10.63	81.50
2240mm	nr	1.45	13.92	61.61	11.33	86.86
2400mm	nr	1.50	14.40	65.69	12.01	92.10
2560mm	nr	1.60	15.36	77.02	13.86	106.24
2720mm	nr	1.80	17.28	81.51	14.82	113.61
2880mm	nr	1.90	18.24	85.99	15.63	119.86
3040mm	nr	2.20	21.12	90.59	16.76	128.47
3200mm	nr	2.50	24.00	94.97	17.85	136.82

APPROXIMATE ESTIMATING

The following composite rates
have been built up from
information given earlier in this
chapter. The rates have been
rounded off for ease of use and
should be helpful when assessing
approximate values of
construction work.

EARTHWORKS	Unit	Approximate prices £
Excavate for trench including supports and sides, level and ram bottom, part return fill and ram, part load into skip, width 600mm, average depth		
by hand		
0.75m	m	20
1.00m	m	25
1.50m	m	38
by machine		
0.75m	m	13
1.00m	m	17
1.50m	m	25

APPROXIMATE ESTIMATING

CONCRETE WORK

		£
Ready mix A plain concrete in foundations, size		
600 x 225mm	m	10
750 x 225mm	m	12
Ready mix B reinforced concrete (1:2:4) in bed with A252 reinforcement, thickness		
100mm	m2	15
150mm	m2	20
Ready mix B reinforced concrete (1:2:4) in wall including wrought formwork both sides and bar reinforcement, thickness		
150mm	m2	88
225mm	m2	98

BRICKWORK

Cavity wall in gauged mortar including forming cavity and wall ties between half brick wall in common and		
100mm blockwork	m2	70
half brick wall in common	m2	64
half brick fair faced wall	m2	66
half brick wall in facings	m2	94
Cavity wall in gauged mortar including forming cavity and wall ties between 100mm blockwork and		
half brick fair faced wall	m2	73
half brick wall in facings	m2	100

ROOFING

Natural Welsh blue slates on battens and felt	m2	64
Marley Modern concrete tiles on battens and felt	m2	20
Woodwool slabs 50mm thick and two layers built up roofing felt	m2	26
Reinforced woodwool slabs 75mm thick and three layers built up roofing felt	m2	50

WOODWORK

19mm thick tongued and grooved boarding to joists, size		
50 x 100mm	m2	28
50 x 125mm	m2	30
75 x 125mm	m2	32
Stud partition, thickness		
75mm	m2	15
100mm	m2	17
Standard flush door plywood both sides 35mm thick, including frame, architrave both sides, fixing only ironmongery, size		
686 x 1981 x 35mm	nr	80
726 x 2040 x 40mm	nr	82
Standard flush door sapele faced both sides 40mm thick, including frame, architrave both sides, fixing only ironmongery, size		
762 x 1981 x 35mm	nr	88
826 x 2040 x 40mm	nr	92

Woodwork (cont'd)

		£
Standard flush door teak faced both sides 40mm thick including frame, architrave both sides, fixing only ironmongery, size		
686 x 1981 x 35mm	nr	110
726 x 2040 x 40mm	nr	112

PLUMBING

Rainwater installation

	£
uPVC, 112mm half round gutter with 75mm downpipe	
terrace house 7m wide with two gutters, outlets, downpipes, offsets and shoes	190
semi-detached house with hip end, three gutters, outlets, downpipes, offsets and shoes	220
semi-detached house with gable end, two gutters, outlets, downpipes, offsets and shoes	190
detached house with gable ends, two gutters, outlets, downpipes, offsets and shoes	204
detached house with hip ends, all round four gutters, outlets, downpipes, offsets and shoes	375
bungalow with gable ends, two gutters, outlets, downpipes, offsets and shoes	290
Cast iron 115mm half round gutter with 75mm downpipes with ears	
terrace house, 7m wide with two gutters, outlets, downpipes, offsets and shoes	370
semi-detached house 8m x 10m with hip end, three gutters, outlets, downpipes, offsets and shoes	435
semi-detached house 8m x 10m with gable end, two gutters, outlets, downpipes, offsets and shoes	375
detached house 9m x 11m with gable ends, two gutters, outlets, downpipes, offsets and shoes	400
detached house 9m x 11m with hip ends all round, four gutters, outlets, downpipes, offsets and shoes	760
bungalow 10m x 10m with gable ends, two gutters, outlets, downpipes, offsets and shoes	530

Plumbing (cont'd)

Hot and cold water installation £
Hot and cold water installation complete; plastic
storage cistern; rising main; down service; gas
multi-point hot water heater supplies to bathroom
and kitchen, sanitary fittings, sink unit and wastes

Semi-detached house (no central heating)
  copper pipework 1340

Hot and cold water installation complete; gas fired
boiler and pump, with plastic storage cistern and
header tank; seven radiators, thermostat, time
clock, rising main, down service, indirect copper
cylinder with immersion heater; hot water supplies
to bathroom and kitchen; sanitary fittings sink
unit and wastes

Semi-detached house
  copper pipework 2700

Add for additional cloakroom on ground floor
with close coupled WC and wash basin 380
Add for four bedroom detached house with
en-suite master bedroom upgraded central
heating system, coloured siphonic sanitary
fittings; shower unit and bidet 2200

Soil and waste installation
Soil and waste installation complete; serving first
floor bathroom with WC, bath, wash basin
  PVC pipework 225
  cast iron pipework with caulked lead joints 500
  cast iron pipework with 'Timesaver' joints 420

Sanitary fittings
Wash basin 560 x 430mm in vitreous china; chromium
plated pillar taps, waste plug and chain on brackets 120

Extra for pedestal 35

Plumbing (cont'd)

	£
Extra for coloured basin 680 x 530mm with pedestal	300

Bath complete with chromium plated pillar taps, waste,
plug and chain, overflow and trap, bath panel to
one side and one end

acrylic white 1700 x 700mm	260
coloured	270
medium quality white 1700 x 700mm	415
'luxury' acrylic coloured 1700 x 850mm	630
steel porcelain enamelled 1700 x 700mm white	240
heavy duty steel porcelain enamelled 1700 x 700mm white	310
Whirlpool bath complete	2000

Washdown WC suite complete with close coupled cistern, seat and overflow pipe, white vitreous china	215
Extra for siphonic type	90
Luxury coloured siphonic WC suite complete	430

FLOOR, WALL AND CEILING FINISHINGS

13mm two coat plaster; browning
and finish to walls; 12.5mm
plasterboard and skim to ceiling;
individual rooms

Floor to ceiling height 2.4m;
floor area

9m2	325
12m2	400
15m2	460

Floor, wall and ceiling finishings (cont'd)

Floor to ceiling height 2.4m; floor area (cont'd)	£
18m2	525
21m2	580
24m2	645
27m2	700
30m2	750

25mm thick granolithic floor;
5mm thick by 150mm high
skirting; individual rooms

Floor area	
9m2	120
12m2	150
15m2	185
18m2	220
21m2	240
24m2	265
27m2	300
30m2	320

150 x 150 x 12.5mm thick
quarry floor tiles bedded in
12mm mortar; 12.5mm thick
x 150mm high skirting:
individual rooms

Floor area	
9m2	300
12m2	390
15m2	470
18m2	550
21m2	650
24m2	725
27m2	750
30m2	885

Floor, wall and ceiling finishings (cont'd)

	£
2.5mm thick Econoflex series 4 vinyl tiles; 100mm high matching vinyl skirting; individual rooms	

Floor area

9m2	160
12m2	190
15m2	230
18m2	260
21m2	295
24m2	325
27m2	360
30m2	380

150 x 150 x 9mm vitrified
ceramic floor tiles; plain; 9mm
thick x 150mm high matching
skirting; individual rooms

Floor area

9m2	370
12m2	480
15m2	580
18m2	680
21m2	790
24m2	900
27m2	1000
30m2	1090

PAINTING

Two coats matt white emulsion paint on

plastered walls	m2	3.00
plastered ceilings	m2	3.50

Prepare prime and apply two undercoats
and one coat gloss on

wood general surfaces	m2	7.20
metal general surfaces	m2	7.70

DRAINAGE

Machine excavation £

Excavate trench, granular bed and
benching, lay 100mm Hepseal pipe,
backfill and remove surplus,
average depth of trench

1.00m	m	25
2.00m	m	40
3.00m	m	50

Excavate trench, granular bed and
benching, lay 150mm Hepseal pipe,
backfill and remove surplus,
average depth of trench

1.00m	m	33
2.00m	m	55
3.00m	m	58

Excavate trench, granular bed and
benching, lay 225mm Hepseal pipe,
backfill and remove surplus,
average depth of trench

1.00m	m	50
2.00m	m	62
3.00m	m	75

Excavate trench, concrete bed and
benching, backfill and remove surplus,
average depth of trench 1m

100mm Hepseal pipe	m	36
150mm Hepseal pipe	m	45
225mm Hepseal pipe	m	70

Manhole including excavation, £
concrete base and benching Class
'B' engineering brick walls,
150mm straight main channel, two
three quarter section bends,
light duty cover and frame, depth
to invert

1.00m	nr	450
1.50m	nr	490
2.00m	nr	525

Chapter 6
Plant and Tool Hire

The prices contained in this chapter are based upon information supplied by HSS Hire Shops Ltd, 25 Willow Lane, Mitcham, Surrey who have over 160 shops throughout the country (see Yellow Pages). The prices exclude VAT and delivery charges.

	First 24 hrs £	Addit 24 hrs £	Per week £
ACCESS AND SUPPORT			
Narrow tower base size 1.3 x 1.5m, height			
2.5m	27.50	13.75	55.00
4.5m	40.50	20.25	81.00
6.5m	53.50	26.75	107.00
8.5m	66.50	33.25	133.00
10.5m	79.50	39.75	159.00
Span tower base size 1.3 x 1.5m or 1.3 x 2.5m, height			
2.5m	27.50	13.75	53.00
4.5m	40.50	20.25	81.00
6.5m	53.50	26.75	107.00
8.5m	66.50	33.25	133.00
10.5m	79.50	39.75	159.00

PLANT AND TOOL HIRE

Access and support (cont'd)	First 24 hrs	Addit 24 hrs	Per week
Alloy ladders			
Double 3.5m extending to 6.2m	9.00	4.50	18.00
5.0m 9.0m	12.50	6.25	25.00
Treble 2.5m 6.0m	9.00	4.50	18.00
3.5m 9.1m	12.50	6.25	25.00
Roof ladders			
Alloy 4.9m, 5.9m and 6.9m	15.00	5.00	25.00
Ladder stay each	5.40	1.80	9.00
Builder's steps			
8 Tread height 1.5m	7.00	3.50	14.00
10 Tread height 2.1m	8.00	4.00	16.00
12 Tread height 2.7m	9.00	4.50	18.00
Steel trestles			
Nos 1-4: 0.5m extending to 2.4m	-	-	3.20
Scaffold boards			
Length 2.4-3.9m	-	-	1.90
Lightweight staging			
Length : 2.4m	8.00	4.00	10.00
3.0m	9.00	4.50	11.25
3.6m	10.00	5.00	12.50
4.2m	11.00	5.50	13.75
4.8m	13.00	6.50	16.25
6.0m	15.00	7.50	18.75
7.2m	20.00	10.00	25.00

PLANT AND TOOL HIRE

BUILDING AND DECORATING	First 24 hrs £	Addit 24 hrs £	Per week £
Metal locator	9.00	3.00	15.00
Cable avoiding tool	32.40	10.81	54.00
Cat signal generator	19.20	6.40	32.00
Steel props Nos 0-4: 1.8m extending to 4.9m (quantity discounts)	-	-	3.20
Jackall prop	9.60	3.20	16.00

Building/decorating tools

	First 24 hrs £	Addit 24 hrs £	Per week £
Wallpaper stripper			
electric	9.60	3.20	16.00
gas	16.80	5.60	28.00
perforator	11.40	3.80	19.00
Tile saw	38.40	11.60	58.00
Hand operated tile cutter	8.40	2.80	14.00
Tile breaker	3.60	1.20	6.00
Damp-proof injection unit	28.80	9.60	48.00
Blowlamp with extension hose	6.60	2.20	11.00
Bolt croppers	7.20	2.40	12.00
Crowbar	3.60	1.20	6.00
Floorboard cramps	4.20	1.40	7.00
G-cramps, sash cramps	3.00	1.00	5.00
Paving mallet	3.60	1.20	6.00
Pickaxe matlock, punner	3.60	1.20	6.00
Shovel/spade	3.60	1.20	6.00
Site tool box	15.60	5.20	26.00
Sledgehammer 3kg and 6.5kg	3.60	1.20	6.00
Spirit level	3.60	1.20	6.00
Tarpaulins	7.20	2.40	12.00
Tyrolean roughcast machine			
manual	9.00	3.00	15.00
electric	16.80	5.60	28.00
Wheelbarrow	3.60	1.20	6.00
White line marker	11.40	3.80	19.00
Workmate	9.00	3.00	15.00

PLANT AND TOOL HIRE

	First 24 hrs £	Addit 24 hrs £	Per week £
Road hazard equipment			
Traffic warning cones (0.5m)	-	-	1.60
Flashing lamp and stand	-	-	5.00
Road signs	-	-	9.50
Road barrier, cone and cap	-	-	8.00
CONCRETING AND COMPACTION			
Concrete mixers			
4/3 ft3 (½ bag), petrol	11.40	3.80	19.00
4/3 ft3 (½ bag), electric	11.40	3.80	19.00
6/4 ft3 diesel	21.00	7.00	35.00
Concrete laying and finishing			
Poker vibrator, electric	27.00	9.00	45.00
Poker vibrator, petrol	31.20	10.40	52.00
Poker vibrator, diesel	31.20	10.40	52.00
Power finishing trowel, petrol	31.20	10.40	52.00
Beam screeds, petrol (per unit)	45.00	15.00	75.00
Needle gun, scaler, air driven	15.60	5.20	26.00
Indent roller	7.20	2.40	12.00
Floor grinder, diesel/electric	54.00	18.00	90.00
Track/floor saw, diesel	39.00	13.00	65.00
Compactors			
Rammer, 2 stroke petrol	30.00	10.00	50.00
Vibrating plate, medium, petrol	25.20	8.40	42.00
Vibrating roller, diesel/petrol	57.00	19.00	95.00
Cowley level	24.00	8.00	40.00
Laser level	57.00	19.00	95.00

PLANT AND TOOL HIRE

	First 24 hrs £	Addit 24 hrs £	Per week £
DRAIN CLEARING/PLUMBING/PUMPING			
Plumber's tools			
Blowtorch	7.20	2.40	12.00
Pipe freezing kit CO2 (gas)	21.00	7.00	24.50
Steel pipe bender, hydraulic	31.20	10.40	52.00
Copper pipe bender			
hand held	9.00	3.00	15.00
large	19.20	6.40	32.00
Steel pipe cutter	6.60	2.20	11.00
Clay pipe cutter	12.00	4.60	20.00
Pipe vice bench	9.60	3.20	16.00
Pipe wrenches			
450mm	4.80	1.60	8.00
600mm	6.60	2.20	11.00
900mm	7.80	2.60	13.00
Chain wrench	9.60	3.20	16.00
Pipe threading			
Electric die stock	39.00	45.50	65.00
Ratchet die stock	15.60	5.20	26.00
Pipe-threading machine ½in-4in electric	72.00	24.00	120.00
Pipe-threading machine ½in-2in electric	54.00	18.00	90.00
Pipe pressure tester	15.60	5.20	26.00
Pipe saw	21.60	7.20	36.00
Water pumps			
Submersible 25mm, electric	12.00	4.00	20.00
Submersible 32mm, electric	18.00	6.00	30.00
Submersible 50mm, petrol	27.60	9.20	46.00
Centrifugal 50mm, petrol	31.20	10.40	52.00
Centrifugal 75mm, petrol	39.00	13.00	65.00
Drain test kit 'U' gauge	6.00	2.00	10.00
Air bag drain stopper	7.20	2.40	12.00

Drain testing/clearing tools	First 24 hrs £	Addit 24 hrs £	Per week £
Drain plugs 100-150mm (per pair)	3.60	1.20	6.00
Drain rods and fittings 9m (per set)	8.40	2.80	14.00
Ropump	8.40	2.80	14.00
Sink cleaner, hand-operated	8.40	2.80	14.00
Drain cleaner, hand-operated	15.00	5.00	25.00
Powered drain cleaner, electric	34.80	11.60	58.00

HEATING, COOLING AND DRYING

Industrial heaters - gas

Plaque heater output, 2,500 Btu	10.80	3.60	18.00
Forced air output, 140,000 Btu	28.80	9.60	48.00
Forced air output, 275,000 Btu	39.00	13.00	65.00

Industrial heaters - paraffin

Forced air, 60,000 Btu	27.00	9.00	45.00
Forced air, 85-100,000 Btu	30.00	10.00	50.00
Forced air, 150,000 Btu	39.00	13.00	65.00

Home/office heaters

Cabinet, gas, 3-16,000 Btu	8.40	2.80	14.00
Electric fan heater, 3kW	6.00	2.00	10.00

Air conditioning units

Air conditioner	57.00	19.00	95.00
Space cooler	51.00	17.00	85.00
Cold air blower	21.00	7.00	35.00

Drying

Building dryer dehumidifier - 300m3	39.00	13.00	65.00
Portable building dryer - 140m3	22.80	7.60	38.00
Portable fume extractor	42.00	14.00	70.00

PLANT AND TOOL HIRE

	First 24 hrs £	Addit 24 hrs £	Per week £
LIGHTING, WELDING AND POWER			
Lighting units			
Gas - large tripod-mounted	12.00	4.00	20.00
Gas - small cylinder-mounted	9.00	3.00	15.00
Electric - tripod-mounted	12.00	4.00	20.00
Plasterers' light - fluorescent	12.00	4.00	20.00
Magnetic 500W flood	11.40	3.80	19.00
Twin 500W flood - 5m tower mast	21.60	7.20	36.00
Festoon lights, industrial (34cm)	12.00	4.00	20.00
Welding			
Arc welder 140/180 amp, 240 volt	19.20	6.40	32.00
MIG welder, 240 volt, medium	25.20	8.40	29.40
No-gas MIG welder	19.20	6.40	32.00
Spot welder, 240 volt (30 amp)	19.20	6.40	32.00
Site welder 20-170 amp, petrol	46.80	15.60	78.00
Welder/generator 300 amp, d.c. silenced	96.00	32.00	160.00
Oxy/acetylene welding kit	28.80	9.60	48.00
Generators (continuous rating)			
3 kVA petrol, 110/240 volt	34.80	11.60	58.00
4 kVA diesel, 110/240 volt	46.80	15.60	78.00
8 kVA diesel, 110/240 volt	72.00	24.00	120.00
15 kVA diesel, 110/240 volt	90.00	30.00	150.00
Transformers			
2.2 kVA	5.70	1.90	9.50
3.0 kVA	8.40	2.80	14.00
5.0 kVA	15.60	5.20	26.00
10 kVA	29.40	9.80	49.00

PLANT AND TOOL HIRE

	First 24 hrs £	Addit 24 hrs £	Per week £
Extension cable 15m cable/drum	4.50	1.50	7.50
Fourway junction box	7.20	2.40	12.00
Power breaker RCD plug	3.00	1.00	5.00

KANGO BREAKING AND DRILLING

Hydraulic breakers

	First 24 hrs £	Addit 24 hrs £	Per week £
Heavy duty, diesel	57.00	19.00	95.00
Medium duty, petrol	48.00	16.00	80.00

Electric hammers

	First 24 hrs £	Addit 24 hrs £	Per week £
Heavy duty breaker (inc. trolley)	33.60	11.20	56.00
Medium duty breaker	18.00	6.00	30.00

Rotary hammers

	First 24 hrs £	Addit 24 hrs £	Per week £
Hilti breaker drill, TE 72	18.00	6.00	30.00
Hilti breaker drill, TE 52	18.00	6.00	30.00
Medium-duty breaker drill	16.20	5.40	27.00
Light duty	16.20	5.40	27.00

Electric drills

	First 24 hrs £	Addit 24 hrs £	Per week £
Two speed percussion drill	8.40	2.80	14.00
Cordless drill	12.00	4.00	20.00
Right angle drill	16.20	5.40	27.00
Four speed drill	16.20	5.40	27.00

Magnetic base drills

	First 24 hrs £	Addit 24 hrs £	Per week £
Magnetic drill stand (drill extra)	30.60	10.20	51.00
Magnetic drill stand c/w drill	48.80	15.60	50.60

PLANT AND TOOL HIRE

	First 24 hrs £	Addit 24 hrs £	Per week £
FIXING, GRINDING AND SANDING			
Fixing tools			
Cartridge hammer	18.00	6.00	30.00
Staple tacker - light duty	6.00	2.00	10.00
Hammer stapler - heavy duty	13.20	4.40	22.00
Nail gun - air operated, automatic	24.00	8.00	40.00
Impact wrench - electric	15.20	5.20	26.00
Screwdriver - electric, auto-feed	16.80	5.60	28.00
Angle grinders			
Angle grinder, 100mm	8.40	2.80	14.00
Angle grinder, 230mm	11.40	3.80	19.00
Angle grinder, 300mm	20.40	6.80	34.00
Sanders			
Belt sanders	15.00	5.00	25.00
Disc sander, 180mm	15.00	5.00	25.00
Orbital sander, industrial	14.40	4.80	24.00
Floor sanders			
Domestic 200mm	24.60	8.20	41.00
Edging sander	21.60	7.20	36.00
Floor sander c/w edging sander	37.50	12.50	62.50
Router			
Laminate trimmer	16.80	5.60	28.00
Power plane	16.80	5.60	28.00

	First 24 hrs £	Addit 24 hrs £	Per week £
SAWING AND CUTTING			
Timber saw benches			
Combination bench, 300mm	51.00	17.00	85.00
Combination bench, 200mm	39.00	13.00	65.00
General-purpose saws			
Crosscut and mitre saw	28.80	9.60	48.00
Flipover mitre saw	39.00	13.00	65.00
Band saw	49.20	16.40	82.00
Reciprocating saw	17.40	5.80	29.00
Jig saw	14.40	4.80	24.00
Circular saw, 230mm blade	15.60	5.20	26.00
Circular saw, 150mm blade	12.60	4.20	21.00
Chain saw with safety kit	39.00	15.00	65.00
Metal and masonry cutting saw			
Masonry sawbench, petrol	54.00	18.00	90.00
Masonry saw bench, electric	45.00	15.00	75.00
Bench top cut-off saw	28.80	9.60	48.00
Portable cut-off saw, electric/petrol	20.40	6.80	34.00
Cut-off saw trolley	6.00	2.00	10.00
Chasing machine	36.00	12.00	60.00
Metal shears	19.20	6.40	32.00
Metal nibblers	19.20	6.40	32.00
Floor saw, 450mm	51.60	17.20	86.00
Floor saw, 350mm	39.00	13.00	65.00
Tile saw	34.80	11.60	58.00
Slab splitter	32.40	10.80	54.00
Block splitter	16.80	5.60	28.00

PLANT AND TOOL HIRE

	First 24 hrs £	Addit 24 hrs £	Per week £
PAINT SPRAYING AND BLASTING			
Spray units			
Portable airless spray	78.00	26.00	130.00
Large volume airless spray	90.00	30.50	150.00
Compressors			
Industrial 9cfm electric	43.20	14.40	72.00
Industrial 15cfm electric/petrol	51.00	17.00	85.00
LIFTING AND MATERIALS HANDLING			
Man lift, height 7.3m	75.00	25.00	125.00
Man lift, height 9.1m	120.00	40.00	200.00
Tirfor winch, TU 16	15.60	5.20	26.00
Tirfor winch, TU 32	19.80	6.60	33.00
Scaffold hoist (200kg)	45.00	15.00	75.00

Chapter 7
General Data

The metric system

Linear

1 centimetre (cm)	= 10 millimetres (mm)
1 decimetre (dm)	= 10 centimetres (cm)
1 metre (m)	= 10 decimetres (dm)
1 kilometre (km)	= 1000 metres (m)

Area

100 sq millimetres (mm^2)	= 1 sq centimetre (cm^2)
100 sq centimetres (cm^2)	= 1 sq decimetre (dm^2)
100 sq decimetres (dm^2)	= 1 sq metre (m^2)

Capacity

1 millilitre (ml)	= 1 cubic centimetre (cm^3)
1 centilitre (cl)	= 10 millilitres (ml)
1 decilitre (dl)	= 10 centilitres (cl)
1 litre (l)	= 10 decilitres (dl)

Weight

1 centigram (cg)	= 10 milligrams (mg)
1 decigram (dg)	= 10 centigrams (cg)
1 gram (g)	= 10 decigrams (dg)
1 decagram (dag)	= 10 grams (g)
1 hectogram (hg)	= 10 decagrams (dag)

GENERAL DATA

Imperial/metric conversions

Linear

1 in = 25.4mm	1mm = 0.03937 in
1 ft = 304.8mm	1cm = 0.3937 in
1 yd = 914.4mm	1dm = 3.397 in
	1m = 39.37 in

Square

1 sq in = 645.16mm^2	1cm^2 = 0.155 sq in
1 sq ft = 0.0929m^2	1m^2 = 10.7639 sq ft
1 sq yd = 0.8361m^2	1m^2 = 1.196 sq yd

Cube

1 cu in = 16.3871cm^3	1cm^3 = 0.061 cu in
1 cu ft = 0.0283m^3	1m^3 = 35.3148 cu ft
1 cu yd = 0.7646m^3	1m^3 = 1.307954 cu yd

Capacity

1 fl oz = 28.4ml	1ml = 0.0352 fl oz
1 pt = 0.568 l	1dl = 3.52 fl oz
1 gallon = 4.546 l	1 litre = 1.7598 pt

Weight

1 oz = 28.35g	1g = 0.035 oz
1 lb = 0.4536kg	1kg = 35.274 oz
1 st = 6.35kg	1t = 2204.6 lb
1 ton = 1.016t	1t = 0.9842 ton

Temperature equivalents

In order to convert Fahrenheit to Celsius deduct 32 and multiply by 5/9. To convert Celsius to Fahrenheit multiply by 9/5 and add 32.

Fahrenheit	Celsius
230	110.0
220	104.4
210	98.9
200	93.3
190	87.8
180	82.2
170	76.7
160	71.1
150	65.6
140	60.0
130	54.4
120	48.9
110	43.3
90	32.2
80	26.7
70	21.1
60	15.6
50	10.0
40	4.4
30	-1.1
20	-6.7
10	-12.2
0	-17.8

BRICKS   Number of bricks per square metre in half brick thick wall in stretcher bond

50 x 102.5 x 215mm	74
65 x 102.5 x 215mm	59
75 x 102.5 x 215mm	52

BLOCKS   Number of blocks per square metre

450 x 225mm	10
450 x 300mm	7
600 x 225mm	7

TIMBER   1 standard = 4.67227 cubic metres

1 cubic metre = 35.3148 cubic feet

10 cubic metres = 2.140 standards

Number of asbestos-free slates per m2

Size (mm)	Lap (mm)	Nr of slates
400 x 200	70	30.0
400 x 200	76	30.9
400 x 200	90	32.3
400 x 240	80	26.1
500 x 250	90	19.5
500 x 250	80	19.1
500 x 250	70	18.6
500 x 250	76	18.9
500 x 250	90	19.5
500 x 250	106	20.5
500 x 250	100	20.0
600 x 300	106	13.6
600 x 300	100	13.4
600 x 300	90	13.1
600 x 300	80	12.9
600 x 300	70	12.7
600 x 350	100	11.5

Blue Welsh slates
Number of slates per m2

Size (mm)	Nr of slates
405 x 205 (16" x 8")	29.59
405 x 255 (16" x 10")	23.75
405 x 305 (16" x 12")	19.00
460 x 230 (18" x 9")	23.00
460 x 255 (18" x 10")	20.37
460 x 305 (18" x 12")	17.00
510 x 255 (20" x 10")	18.02
510 x 305 (20" x 12")	15.00
560 x 280 (22" x 11")	14.81
560 x 305 (22" x 12")	14.00
610 x 305 (24" x 12")	12.27

GENERAL DATA

Westmorland green slates
1 ton (Imperial) standard quality covers approximately 18-20m2

1 ton (Imperial) Peggies covers approximately 15-16m2

Marley tiles (100mm gauge)

Type	Nr/m2
Plain	60.0
Feature	56.0
Ludlow Plus	17.4
Anglia Plus	17.3
Ludlow Major	10.7
Mendip	10.6
Double Roman	10.4
Modern	10.8
Wessex	11.0
Bold Roll	10.6
Monarch	14.5

Lead

Code	kg/m2
3	14.97
4	20.41
5	25.40
6	30.05
7	36.72
8	40.26

Number of tiles per square metre

150 x 150mm	44
100 x 200mm	50
200 x 200mm	25
250 x 125mm	32
230 x 230mm	19

Coverage of plasters

Carlite premixed browning	m2/1000kg
11mm floating coat	130-150

GENERAL DATA

Metal lathing   11mm pricking up and floating   60-70

Bonding coat   8mm floating coat on concrete   145-155

           11mm floating coat on brickwork and
           blockwork   100-110

           8mm floating coat on plasterboard   150-165

Finish   2mm finishing coat on floating coat   410-500

Thistle final coat plasters
Thistle finish   2mm finishing coat on sanded undercoat 350-450

Thistle board
finish   5mm finishing in two coats   160-170

Thistle renovating plasters
Thistle
undercoat   11mm thick   120

Thistle finish   2mm thick   380-420

Sirapite B plasters
Sirapite B   3mm finishing coat on sanded undercoat   250-270

Mortar mixes for plaster used in the manual
Cement mortar   1:3   0.48 tonnes cement/m3
                           1.45 tonnes sand/m3

                 1:4   0.36 tonnes cement/m3
                           1.45 tonnes sand/m3

Cement lime mortar 1:1:6   0.22 tonnes cement/m3
                            0.11 tonnes lime/m3
                            1.45 tonnes sand/m3

                 1:2:9   0.16 tonnes cement/m3
                            0.14 tonnes lime/m3
                            1.45 tonnes sand/m3

Index

MINOR WORKS, ALTERATIONS, REPAIRS AND MAINTENANCE

Other titles from E & FN Spon

Building Regulations Explained
1992 Revision
J. Stephenson

CESMM3 Explained
B. Spain and L. Morley

Commercial Estimator
Marshall & Swift

Residential Estimator
Marshall & Swift

Construction Contracts
Law and Management
J. Murdoch and W. Hughes

Construction Tendering and
Estimating
J. I. W. Bentley

Effective Speaking
Communicating in speech
C. Turk

Effective Writing
Improving scientific, technical and
business communication
C. Turk and J. Kirkman

Estimating Checklists for Capital
Projects
2nd edition
The Association of Cost Engineers

Good Style
Writing for science and technology
J. Kirkman

Housing Defects Reference Manual
The Building Research Establishment
Defect Action Sheets
Building Research Establishment

Post-Construction Liability and
Insurance
Edited by J. Knocke

The Presentation and Settlement of
Contractors' Claims
G. Trickey

Project Budgeting for Buildings
D. Parker and A. Dell'Isola

Project Management Demystified
Today's tools and techniques
G. Reiss

Spon's Budget Estimating Handbook
2nd edition
Spain and Partners

Spon's Building Costs Guide for
Educational Premises
Tweeds

Spon's Construction Cost and Prices
Indices Handbook
M. Flemming and B. Tysoe

Spon's European Construction Costs
Handbook
Davis Langdon & Everest

Standard Method of Specifying for
Minor Works
3rd edition
L. Gardiner

Understanding JCT Building Contracts
3rd edition
D. M. Chappell

Write in Style
A guide to good English
R. Palmer

For more information on these and other titles please contact:
The Promotion Department, E & FN Spon,
2–6 Boundary Row, London, SE1 8HN.
Telephone 071-522 9966

SPON'S
CONTRACTORS' HANDBOOK

MINOR WORKS, ALTERATIONS, REPAIRS AND MAINTENANCE

1995

Tweeds

CHARTERED QUANTITY SURVEYORS,
COST ENGINEERS, CONSTRUCTION ECONOMISTS

E & FN SPON
An Imprint of Chapman & Hall

London · Glasgow · Weinheim · New York · Tokyo · Melbourne · Madras

**Published by E & FN Spon, an imprint of Chapman & Hall,
2–6 Boundary Row, London SE1 8HN**

Chapman & Hall, 2–6 Boundary Row, London SE1 8HN, UK

Blackie Academic & Professional, Wester Cleddens Road, Bishopbriggs,
Glasgow G64 2NZ, UK

Chapman & Hall GmbH, Pappelallee 3, 69469 Weinheim, Germany

Chapman & Hall USA, One Penn Plaza, 41st Floor, New York NY 10119,
USA

Chapman & Hall Japan, ITP-Japan, Kyowa Building, 3F, 2-2-1 Hirakawacho,
Chiyoda-ku, Tokyo 102, Japan

Chapman & Hall Australia, Thomas Nelson Australia, 102 Dodds Street,
South Melbourne, Victoria 3205, Australia

Chapman & Hall India, R. Seshadri, 32 Second Main Road, CIT East,
Madras 600 035, India

First edition 1986
Seventh edition 1994

© 1994 E & FN Spon

Printed in Great Britain by St Edmundsbury Press
Bury St Edmunds, Suffolk

ISBN 0 419 18540 2

A catalogue record for this book is available from the British Library

∞ Printed on permanent acid-free text paper, manufactured in accordance with
ANSI/NISO Z39.48-1992 and ANSI/NISO Z39.48-1984 (Permanence of
Paper).